Praise for Peter Maass's

CRUDE WORLD

"Sharp. . . . Maass succee   ostly ignored or misunderstood in the developed world." —*Financial Times*

"Riveting and illuminating. . . . The book is not about oil policy or the energy crisis, at least not primarily; it is a moral reckoning with basic instincts." —*The Nation*

"A nice primer on the history of oil and its geopolitical ramifications." —*Houston Chronicle*

"A fascinating, nightmarish journey to the far end of the pipeline. If you want to know the true cost of America's oil addiction—and even if you don't—you should read this book." —Elizabeth Kolbert, author of *Field Notes from a Catastrophe*

"A spare, engaging work of reporting." —Robert D. Kaplan, *The Wall Street Journal*

"A disturbing catalogue of the underside of the international oil industry. . . . This is not a dispassionate exploration of sticky business issues; it's an indictment and a conviction wrapped in one." —*The Washington Post*

"The strength of *Crude World*, filled with vivid reporting, is that it leaves you no option but to care." —*The Observer* (London)

"A fascinating if alarming book. . . . Nightmare scenarios and grim reality checks are scary stuff, but Maass delivers the bad news in prose that teems with keen observations and well-reported, unforgettable details." —*Richmond Times-Dispatch*

"*Crude World* gets its energy from Peter Maass's exhaustive investigation and firsthand experience and results in an illuminating narrative of the true impact of the global dependence on oil. . . . Essential reading for these times and for anyone interested in making the right decisions about our energy future."
—Robert Redford

"What's compelling [about *Crude World*] is the reporting Maass has done from places ruined by dependence on a substance that, he argues convincingly, kills more than it liberates."
—*The Boston Globe*

"Unflinching. . . . Engrossing. . . . Equal parts *Heart of Darkness* and *Mad Max*."
—*The Washington Monthly*

"With the clarity of a hard-boiled investigator and the grace of a fine writer, Peter Maass reveals how oil has cursed the countries that possess it, corrupted those who want it, and wrought havoc on a world addicted to it. Brilliant and compelling."
—Robert B. Reich, author of *Supercapitalism*

"A lucid account that places readers in the big picture. . . . [Maass's] truths are painful and unavoidable. . . . He has written a very brave book."
—*Fredericksburg Free Lance-Star*

"Well argued and illustrated. . . . [Maass's] descriptions of the rape of countries as diverse as Nigeria and Ecuador are all the more devastating for being unpreachy."
—*The Telegraph* (London)

"Getting off oil is a great idea for a lot of reasons, like saving the planet's climate. But Peter Maass gives us another set of bonuses. If you think drug dealing is a dirty business, then meet the biggest drug of all."
—Bill McKibben, author of *Deep Economy*

Peter Maass

CRUDE WORLD

Peter Maass is a contributing writer to *The New York Times Magazine* and has reported from the Middle East, Asia, South America, and Africa. He has written as well for *The New Yorker*, *The Atlantic Monthly*, *The Washington Post*, and *Slate*. Maass is the author of *Love Thy Neighbor: A Story of War*, which chronicled the Bosnian war and won prizes from the Overseas Press Club and the *Los Angeles Times*. He lives in New York City.

www.petermaass.com

ALSO BY PETER MAASS

Love Thy Neighbor: A Story of War

CRUDE WORLD

The Violent Twilight of Oil

Peter Maass

The Library of Congress has cataloged the Knopf edition as follows:
Maass, Peter.
Crude world : the violent twilight of oil / Peter Maass.—1st ed.
p. cm.
Includes bibliographical references and index.
1. Petroleum industry and trade—History. I. Title.
HD9560.5.M23 2009
338.2'7282—dc22   2009012303

For Alissa

The meek shall inherit the Earth, but not the mineral rights.

J. PAUL GETTY

CONTENTS

CRUDE WORLD

Introduction

Oil was everywhere. It was buried under deserts to the north and south of me, in delectable reservoirs measured in the billions of barrels. Inches away, it fueled the engine that spewed warm exhaust in my face, making my skin smell of diesel. Oil even permeated the minds of the men who jostled around me, talking and shouting in a plaza where the fountains had no water and the grass was not green and the violence was panoramic.

One of the ironies of oil-rich countries is that most are not rich, that their oil brings trouble rather than prosperity. From where I stood, I could literally see this. The forlorn plaza, whose name, translated into English, meant "Paradise," had once been a pleasant spot to pause amid sculptures of martial greatness; slats of shade were cast by concrete columns arrayed around the statue of a leader whose right arm pointed ahead. Repose had been enjoyed on this patch of earth, but it would not be on this day.

It was April 9, 2003. The invasion of Iraq had culminated in Baghdad with American marines taking control of Firdos Square and wrapping a metal chain around the statue of Saddam Hussein. Beamed across the globe, a series of images would become iconic: a broad-shouldered weight lifter pounding the base of the statue with a sledgehammer, a young marine swaddling an American flag over the statue's head, and the edifice tumbling down and being swarmed by a handful of Iraqis, some using their shoes to slap the dictator's inert face.

The diesel fumes in my face came from the marine vehicle pulling down the statue. I had jumped onto its flank and shouted questions at its plucked-from-obscurity crew. Corporal Edward Chin, who put the flag on Saddam's head, was like a kid at his first carnival. "Crazy," he yelled. "Just crazy." I trotted behind the Iraqis dragging Saddam's bronze head around the square and stood beside one of the parched fountains with Lieutenant Colonel Bryan McCoy, who commanded the troops; he carefully described the statue's toppling as "the Iraqi people's idea." For weeks I had followed these marines to Baghdad in an odyssey of conquest that left a trail of bodies on the deserts and roads behind us, the intended as well as the collateral damage of warfare.

Had Americans marched on Baghdad for the sake of oil rather than democracy or weapons of mass destruction? A young European woman shouting at an exhausted marine in the square-turned-into-theater-of-war believed the answer was yes. She had stayed in Baghdad during the invasion as a "human shield," and she called the marine an imperialist and a warmonger and nearly spat on his flak jacket. It seemed, in her confident fury, that she might strike him. The marine stared back at an assault no profane drill sergeant had prepared him for. Words could not settle this argument, and neither of the protagonists seemed to know what to do or why, really, they were there.

As Saddam's neck was fitted with its metal noose, I chatted with Samir, an Iraqi who said he had lived in America for several years. "I feel good," he told me. "Free at last, free at last." An hour later, someone shouted that Samir was a spy for the now-fugitive Saddam, and several men set upon him, as hounds to a fox. "Kill him!" they shouted, landing their blows. Bloodied and terrified, Samir was saved by a marine waving his Beretta at the mob and yelling, as though trying to persuade his unsure self, "I can't allow this to happen!"

The marine hadn't a clue who Samir was. Neither did I.

Baghdad, and the role oil played in the motives of its invaders, was not coming into focus in the first hours of what was supposed to be a clarifying triumph. True understanding, like oil itself, was buried somewhere. Donald Rumsfeld, who was the secretary of defense at the time, said the invasion had "nothing to do with oil, literally nothing," and I knew this was false. Yet the contrary idea—Iraq invaded only for oil—was not convincing. As the sun went down on Firdos Square, I could not articulate what we do for oil and what oil does for us.

Across the world, oil is invoked as a machine of destiny. Oil will make you rich, oil will make you poor, oil will bring war, oil will deliver peace, oil will define our world as much as the glaciers did in the Ice Age. If the inner workings of this machine are understood, perhaps an order will be revealed in the world's disorder. But we are not talking about a contraption of pistons and gears that can be schematized in a precise way. We are presented with too many moving parts, too many clues that defy easy assemblage.

Here's one of them: I lived in Asia for several years and wondered, if oil was such a blessing to countries possessing it, how South Korea, which has no oil, became an economic tiger, as well as Japan, whose oil reserves are minuscule. Their prosperity in the last fifty years was in contrast to oil exporters like Iraq, Iran and Nigeria, which did not have the profiles of winners. Among their humiliations, the most absurd is that they have shortages of gasoline. They are examples of what economists call the "resource curse," which posits that countries dependent on resource exports—especially oil but also natural gas, diamonds and other minerals—are susceptible to lower growth, higher corruption, less freedom and more warfare. As the graffiti I saw on a pipeline in Ecuador's Amazon region stated, "Más Petróleo = Más Pobreza"— more oil equals more poverty. This seems counterintuitive, especially if prices rise beyond $100 a barrel, but prices vary over time and windfalls tend to be squandered by governments prone to corruption and ineptness, which is to say, most governments. Petrodollars can make them richer but not more honest, efficient or intelligent; usually the opposite occurs.

I was flummoxed by the differing fates of countries that lived off oil. In one, there was invasion. In another, poverty. Nearby, fundamentalism. Not far away, empire. Across an ocean, pollution. In the distance, anarchy. Over the border, civil war. There was prosperity and a measure of freedom in just a few cases, usually in tiny nations with disproportionately large reserves, such as Kuwait, Brunei and the United Arab Emirates. Norway was the outlier of outliers—a country with abundant oil and gas as well as a vibrant economy and an open political system. If oil was a drug, as has been said, it was a volatile one that caused disparately negative reactions in most countries that filled their treasuries with its wages. Why did this happen?

There was another set of questions on the other side of the equation, because oil distorted the destiny of countries that were its consumers. If the fields around Kirkuk and Basra were the reasons Iraq was invaded when a barrel cost just $30 in 2003, it would seem possible that in the years ahead the world's largest consumers of energy might shed more blood or offer other sacrifices of a political or moral nature. And that may not be the worst of it. In this era of climate change, it does not matter which side of the supply-demand continuum you live on; no one escapes the perils of a planet warmed by burning fossil fuels. There is much to learn about oil's destructive hold on us.

But how do you coax secrets from a liquid? To know a person, you talk to him. To know a country, you visit it. To know a religion, you study sacred texts. Oil defies these norms of interrogation. It is a commodity that is extracted, refined, shipped and poured into your gas tank with few people seeing it. It has no voice, body, army or dogma of its own. It is invisible most of the time, but, like gravity, it influences everything we do.

My journey into oil was stirred by a desire to learn more about the reasons for global conflict and poverty and their possible solutions. I knew that the war zones I'd visited since the 1980s were consequences rather than explanations. My inquiries into oil began just before the 9/11 era, when prices were low, climate change was a maligned hypothesis and Osama bin Laden was unknown in most households. In those days, when I searched Amazon.com for books on oil, the proffered list included more tomes on salad dressing and aromatherapy than on the liquid that was the oxygen of the global economy. Those were quiet times, and they did not last.

The countries I visited specifically for this book include Saudi Arabia, Russia, Kuwait, Iraq, Nigeria, Venezuela, Ecuador, Azerbaijan, Pakistan, Equatorial Guinea and Afghanistan. These were in addition to oilcentric nations I had reported from previously, including China, Japan, Sudan, Kazakhstan, Britain and Norway. I have spoken with oilmen in Houston, princes in Riyadh, lobbyists in Washington, roughnecks in Baku, warlords in Nigeria, leftists in Caracas, billionaires in Moscow, environmentalists in Quito, generals in Baghdad, traders in Manhattan, wildcatters in Midland, Texas, and diplomats in London. If you have conversations with people such as these, the topics you discuss include not just politics and economics but history, geology, geography, chemistry, engineering, physics, climatology, ecology, accounting, law, corruption, culture, psychology, anthropology, greed, envy, disease, ego and fear. The world of oil is an intellectual as much as a physical space.

My travels coincided with a spectacular rise in prices, followed by a spectacular fall in 2008, as well as spectacular wars in Iraq and Afghanistan; my timing could not have been better or more complex. Everyone was talking about oil, in whispers and shouts that melted into nonsense at times, into violence on other occasions. I often wished for a simple narrative that would tie everything together in a neat package, a unified theory of oil and life. A false epiphany came in Baghdad.

After the statue of Saddam was toppled, I found a room at the Palestine Hotel and watched from my balcony as the sun went down on a city in flames. The looting had begun. The next morning, I drove with two friends into the free-for-all. Around me, there was pillaging—of offices, stores, banks, warehouses, hospitals, police stations, army bases, municipal buildings, pharmacies and hair salons. The streets filled with cars, tractors and donkey carts piled high with loot. For the thoroughly bereft who lacked even a wheelbarrow, wheeled office chairs were used to move whatever could fit onto the seats—a television, a computer, a tire. Even the National Museum was a target; the treasures of Mesopotamia were being plundered as I drove by.

There was an oasis amid this anarchy. American soldiers had piled sandbags around the Ministry of Oil and were patrolling its perimeter, their .50-caliber machine guns establishing havoc's demarcation line.

The ministry was being spared desecration by mob. This stood out, because American forces protected little else. In the weeks and months ahead, as chaos continued elsewhere in Baghdad and in the country at large, the cordon around the nine-story, sand-colored ministry was unstinting. At the time, the reason for its careful preservation seemed obvious.

Roland Barthes described photographs as violent because they "fill the sight by force," pushing out other views. This process occurred in Baghdad, where many of us fixated on the diametrical fates of the Oil Ministry and the National Museum; they seemed to tell us everything we needed to know about war and oil and the priorities of an invader that appreciated cheap gasoline. But there was far, far more to learn about the black magic shaping our crude world.

For a man warning of a coming apocalypse, Matthew Simmons was surprisingly cheerful.

Simmons is the prosperous founder of a niche investment firm that specializes in mergers and acquisitions in the energy sector. A graduate of Harvard Business School, a member of the Council on Foreign Relations, a frequent flier on corporate jets, an adviser to George W. Bush during his first presidential campaign, and a best-selling author, Simmons would be a card-carrying multimillionaire member of the global oil *nomenklatura*, if cards were issued for such things. But his wonderful life has a cloud, which consists of his belief that the American dream and the world as we know it are on the verge of falling apart.

"If it turns out I am wrong, I will be the happiest guy in the world," Simmons told me. Despite his dire prediction, he retained a mischievous demeanor. We met during a trip he made to Manhattan to talk to a group of oil-shipping executives. The impression he gave was of an enthusiastic inventor sharing a discovery (the coming apocalypse) that had taken him by surprise. He had a wide-eyed wonder in his regard, as if a bit of mystery could be found in everything that caught his eye. Whether he was saying the sky was blue or that it was falling, he emitted a happy wink.

His journey into the apocalypse began innocently enough. In 2003, Simmons visited Saudi Arabia on a government-sponsored tour for business executives. The group was presented with the usual dog and pony show, but instead of being impressed with the size and expertise of the kingdom's oil industry, Simmons became perplexed. A senior manager at Saudi Aramco, the state-owned oil company, told the visitors that "fuzzy logic" was used to estimate the oil that could be recovered from the country's reservoirs. Simmons wondered what could be fuzzy about the contents of an oil reservoir. He realized that Aramco, despite its assurances of bottomless supplies, might not know how much oil remained.

He returned home with an intellectual itch. Saudi Arabia was one of the charter members of OPEC, an organization founded in 1960 to coordinate the policies of the major oil producers. Like every OPEC member, Saudi Arabia provides only general numbers about its output and reserves. It does not release details of how much oil is extracted from each reservoir or what methods are used to extract that oil or how much water emerges with the oil, and it does not permit outside audits or inspections. The condition of Saudi fields—like the condition of almost all OPEC fields—is a closely guarded secret. That's partly because the cartel's production quotas, which keep oil off the market to

In the Shaybah oil field in Saudi Arabia

avoid gluts that drive down prices, are based on each country's reserves. The larger an OPEC member's reserves, the higher its quota. Most, if not all, OPEC members have exaggerated the size of their reserves to have the largest possible quota and thus the largest possible revenues.

During the decades of low prices and overabundant supplies, oil-men like Simmons did not care about OPEC's fudged numbers. Whether or not OPEC reserves were hyped, whether or not the cartel's fields were nearing a midlife crisis, plenty of crude was coming out of the ground. Only a boycott or a war could squeeze supplies, it seemed, and only for a short time. But while visiting Saudi Arabia in 2003, when the price of a barrel was around $30 and rising, thanks in part to China's emergence as a voracious importer, Simmons asked himself whether something might be amiss. Perhaps the world didn't have as much oil as was thought.

As a teenager, Simmons had been on a high school debate team that reached the state championships, and from that experience he'd learned the importance of possessing information, of knowing his subject. He has never been a fan of fuzzy logic or fuzzy anything; he likes hard data. He wanted to know, to *truly* know, what was happening in Saudi fields—how many wells were drilled into them, how much oil each well produced, which recovery methods were used and the trend line of output since the wells had been opened.

It was a mystery crafted for his curious mind. After returning home and immersing himself in the minutiae of oil geology, Simmons realized that uncollated data might be found in the neglected research of the Society of Petroleum Engineers. Oil engineers, like most professionals, have regular conferences at which they present obscure papers that delve into the work they do. Technical and detailed, the papers are debated at the conferences, published by the SPE in a little-read journal, and then forgotten.

Simmons realized that the answers were hiding in plain sight. He found more than two hundred papers that no one had pulled together into a meta-analysis. Though they covered only portions of the kingdom's wells and some of them were several decades old, they contained useful data about the condition of the fields. Two months before we met in New York City, Simmons had shocked the oil world by publishing a book, *Twilight in the Desert*, in which he said:

> The geological phenomena and natural driving forces that created the Saudi oil miracle are conspiring now in normal and predictable ways to bring it to its conclusion, in a time frame potentially far shorter than officialdom would have us believe. . . . Saudi Arabia clearly seems to be nearing or at its peak output and cannot materially grow its oil production.

The denunciations arrived by the supertanker. Saudi officials described Simmons as deluded. Daniel Yergin, the closest thing the oil world has to a guru, treated Simmons as a naïve, overcaffeinated student jumping out of his seat to insist that the professor is wrong; the professor patted the student on the shoulder and kindly asked him to sit back down.

Between Simmons's trip to Saudi Arabia and the publication of his book two years later, in 2005, the price of oil had nearly doubled. When we met at that time, Simmons had a look of mischief on his face, even though he was saying the world might be heading toward disaster.

At times of economic expansion, why does oil cost so much? Before the onset of the global economic crisis in the fall of 2008, a barrel fetched $147. The price is expected to return to triple digits when the recession ends. Just how high will it go? That leads to the essential question of how much oil is left in the world.

If you listen to Ali al-Naimi, the Saudi oil minister, you will hear about infrastructure. He tells audiences there are not enough wells, pipelines, or refineries to extract and distribute the quantities of oil the United States, China and the European Union require in boom times. He says that when there's a price spike, we should cut back on driving or buy a smaller car and wait a few years, and soon the world will again have enough petroleum. Yergin likes to emphasize political problems— if the output of Venezuela, Iraq, Iran and Nigeria were not reduced by instability at times of high prices, Americans would hardly feel a pinch when they filled up their tanks. Yergin and Naimi bemoaned the speculators who traded oil futures not to actually take delivery of the stuff one day but to profit from rising prices. There is an echo of truth in what Yergin and Naimi said—particularly on the role of speculators, who indeed poured money into oil markets and thereby accentuated the rise in prices—but most of it is willful or wishful deception.

Simmons reintroduced the world to a phenomenon known as "peak oil." Its statistical foundation was discovered long ago by M. King Hubbert, a Shell geologist who predicted in 1956 that America's oil output (not including Alaska) would peak by 1970. Hubbert's prediction, derided when he made it, turned out in broad terms to be accurate. His forecast was based on the production trends of reservoirs he studied. He noticed a bell curve in which output rose until the reservoir was half empty, and then output dropped as quickly (or slowly) as it had risen. At the halfway point, reservoirs continued to yield oil, but the amounts slipped year by year because the fields had lost what was, in essence, their geological vigor. Think of an oil field as a runner reaching top speed. Just as the runner cannot increase her pace beyond a certain point, and must slow down after reaching top speed, so does the output of an oil field reach its peak and then decline.

Simmons caused a sensation with his assertion that Saudi fields were peaking. And he went further. Because most of the world's "supergiant" fields were discovered more than three decades ago (some more than seven decades ago), Simmons noted, they were all losing vigor. Burgan, in Kuwait, was declining; Cantarell, in Mexico, was collapsing; Alaska's Prudhoe Bay was limping; Norway and Britain's mother lode in the North Sea was gasping. If you look at those depleting fields, it is no surprise that oil becomes triple-digit expensive when the global economy expands and requires greater amounts of petroleum. Our requirements, in times of expansion, are too vast to be sated by younger but smaller fields. Nor can we squeeze large amounts from new sources of "unconventional" crude, such as tar sands in Canada and heavy oil in Venezuela; these forms of protocrude are difficult to convert into consumable oil and inflict extraordinary damage on the environment.

The pinch of $147-a-barrel oil in 2008 was just a foretaste of what awaits us. To understand why this is true, it is necessary to follow the oil to Saudi Arabia.

The largest oil terminal in the world, Ras Tanura, is located on the eastern coast of Saudi Arabia, along the Persian Gulf. From Ras Tanura's control tower, I saw the classic totems of oil's dominion: supertankers coming and going on an emerald sea, row upon row of storage tanks, and miles and miles of pipes twisting around and between refinery stacks. The industrial panorama was part M. C. Escher, part Rube Goldberg drawing. Ras Tanura is the funnel through which nearly 10 percent of the world's supply of petroleum flows, and its command center had the feel of mission control at NASA—banks of touch-screen monitors, a quiet hum of high-tech efficiency. Standing in the control tower, I was surrounded by more than 50 million barrels of ready-for-export oil, yet not a drop could be seen.

The oil was there, of course. In a technological sleight of hand, oil can be extracted from the deserts of Arabia, processed to eliminate

At the Ras Tanura export terminal in Saudi Arabia

water and natural gas, sent through pipelines to a terminal on the gulf, loaded onto a supertanker and shipped to a port thousands of miles away, then run through a refinery and poured into a tanker truck that delivers it to a suburban gas station, where it is pumped into an SUV— all without anyone actually glimpsing the stuff. And so long as there is enough oil to fuel the global economy, it is not only out of sight but out of mind, at least for most consumers.

I visited Ras Tanura because oil was not out of my mind. I am old enough to remember, as a young boy, the round-the-block lines at American gas stations in 1973. Responding to American support for Israel in the Yom Kippur War, Arab members of OPEC boycotted shipments to the United States, creating an artificial shortfall across the globe that caused a quadrupling of world prices, from $3 a barrel to $12. The world awoke from decades of petroslumber. By 1979, with the Iranian revolution, a barrel reached nearly $40. Eventually, as relative peace returned to the Middle East, prices drifted downward until, in the wake of the Asian financial crisis, a barrel cost only $12 in 1999. OPEC, slow to react to the Asian crisis, belatedly tightened its quotas and, as the world economy expanded again, prices moved upward once more.

With the global economy in recession in late 2008, oil prices collapsed from the highs reached in the boom times of just a few months earlier. Recovery and triple-digit terrain are inevitably linked, with prices destined to shoot higher at the smallest hiccups—the world, in other words, that existed when I visited Ras Tanura, where Aref al-Ali, my escort from Saudi Aramco, gestured at the storage tanks around us. "One mistake at Ras Tanura, and the price of oil will go up," he noted. There was pride in his voice, but also fear.

The port was a fortress. Its entrances had an array of security gates and bomb barriers to prevent terrorists from cutting off the black oxygen the modern world depends on. Before reaching Ras Tanura, we had to pass through several Saudi Arabian National Guard checkpoints on the highway from Dhahran. Even Ali, who worked in Aramco's headquarters in Dhahran, needed special permission to enter Ras Tanura. The House of Saud was concerned about the havoc that could be wrought by a speeding zealot with fifty pounds of TNT in the trunk of his car. But the Saudis had even greater worries.

Two things can ruin Saudi aspirations for another fifty years of financial windfalls. Global warming has fomented worldwide efforts to discourage the use of fossil fuels and develop alternative forms of energy. For Saudis, this is akin to turning their gold into dust. But it is not just a warming planet that is scaring their customers. Rising prices drive them away, too. The more oil costs, the more incentive consumers have to use less of it—and that explains why Americans, when the price of a gallon reached $4, finally began to cut back on their driving. The last thing Saudi Arabia wants is for its clients to conclude that they must find other energy sources because oil is running out and prices for it will only get higher.

National leaders, politicians and economists are not the only ones confounded by oil. Geologists are tricked by the substance, too. Oil cannot be inventoried, like timber in a wilderness. It is underground, unseen by engineers, who can, at best, make educated guesses about how much is there. What's known is that as much as half of the world's proved reserves of conventional oil have been consumed, but at least a trillion barrels remain. (The large reserves of "unconventional" oil in Canada and Venezuela are unlikely to provide more than modest new supplies, due to the difficulty of turning them into usable oil.) A trillion barrels of yet-to-be-recovered oil is quite a lot, but there's a rub. The notion of reservoirs as underground lakes, from which wells extract oil like straws sucking a milkshake, is incorrect. Oil exists in drops squeezed inside rocks such as sandstone. A new reservoir may contain sufficient pressure to make these drops flow to the surface in a gusher, but after a while—usually within a few years and often sooner—natural pressure lets up and is no longer strong enough to push oil to the surface. At that point, "secondary" recovery efforts are begun, like pumping water or gas into the reservoirs to increase the pressure.

This process is unpredictable because reservoirs are fickle. If too much oil is extracted too quickly, or if the wrong types or amounts of secondary efforts are used, the quantity of oil that can be recovered from a field can be greatly reduced; this is known as "damaging a reservoir." It does not matter how many wells are drilled—the field will not yield more oil and, in fact, its output might collapse. This is what Hubbert realized in 1956. A modern example is Oman. In 2001, its daily output reached nearly 960,000 barrels before suddenly falling. The country went on a multibillion-dollar program to modernize its recovery techniques; in addition to injecting advanced detergents and polymers into the reservoirs, engineers lit fires underground to force the oil out (a trick known as in situ combustion). Nonetheless, Oman's production has fallen to less than 800,000 barrels a day. Herman Franssen, a consultant who worked there for a decade, sees a lesson for nations that try to sustain high levels of output. "They used all these new technologies," he told me, "but they haven't been able to stop the decline."

Saudi Arabia may have enough oil to last for generations, but that is not the issue. Crunch time comes long before the last drop of oil is sucked from the Arabian desert. It begins when producers are unable to increase their output. If we do not know when that moment will arrive—and it may arrive any day now—we cannot know when to begin preparing for it, so as to soften its impact. The blow may come like a sledgehammer from the darkness. That's why the debate over peak oil is not just about numbers. It is about the future.

Saudi Arabia possesses 21 percent of the world's conventional reserves. The kingdom has 264 billion barrels, almost twice as much as the runner-up, Iran. Every day, the Saudis provide about 9 million barrels of the approximately 85 million barrels the world consumes. New fields are discovered now and then in other countries, but they tend to offer only small increments. The much-contested reserves in Alaska's Arctic National Wildlife Refuge probably amount to only 10 billion barrels. When the world needs more oil, it has little choice but to call on Riyadh.

Before visiting the kingdom I tried to top off my understanding of its earthly treasure by going to Washington, D.C., to hear a speech that Oil Minister Naimi was delivering at a conference just a few blocks from the White House. Naimi was the star attraction at a gathering of the American petropolitical nexus. Samuel Bodman, the U.S. energy secretary at the time, was on the dais next to him. David O'Reilly, chairman and chief executive of Chevron, was waiting in the wings. The moderator was an éminence grise of the oil world, James Schlesinger, a former energy secretary, defense secretary and CIA director.

"I want to assure you here today that Saudi Arabia's reserves are plentiful, and we stand ready to increase output as the market dictates," said Naimi, dressed in a gray business suit. "I am quite bullish on technology as the key to our energy future. Technological innovation will allow us to find and extract more oil around the world." He described the task of increasing output as just "a question of investment" in new wells and pipelines, and he noted that consuming nations need to build new refineries to process increased supplies of crude. "There is absolutely no lack of resources worldwide," he reiterated.

Naimi's spokesman, Ibrahim al-Muhanna, told me that his boss was too busy for an interview but might have time in Riyadh. When I arrived in the kingdom a few weeks later, Muhanna said that not only would an interview with Naimi be impossible, but I could not talk to

Ali al-Naimi, Saudi minister of oil

anyone at the ministry. It was as though my queries constituted an unfit challenge to the kingdom's geological manhood. At the last minute I was allowed to see Ras Tanura, and I was encouraged to visit Aramco's oil museum in Dhahran, but that is something a Saudi schoolchild can do on a field trip. After a volume of phone calls, Muhanna finally agreed to see me, but not at the ministry. We got together in the lobby of my hotel. He began by noting that the Saudis are no different from other oil producers who refuse to divulge production data.

"They will not tell you," he said. "Nobody will. And that is not going to change." Referring to the fact that Saudi Arabia is often called the central bank of oil, he added, "If an outsider goes to the Fed and asks, 'How much money do you have?' they will tell you. If you say, 'Can I come and count it?' they will not let you. This applies to oil companies and oil countries."

Muhanna was aware of the absurdity of what he was saying, because there is no dispute about the financial reserves of the United States, and in our digitized world monetary reserves are tracked by computers rather than stored in warehouses filled with hundred-dollar bills from floor to ceiling. Muhanna was just doing his job, which in this case consisted of throwing rhetorical dust in my eyes. I responded by mentioning that outsiders remained unconvinced by the Saudis' "trust us" stance, in light of the legendary cheating in OPEC and in the industry. I noted that Royal Dutch/Shell had just admitted that it had overstated its reserves by nearly 24 percent. The quality of slipperiness would seem to apply to more than the physical properties of oil. Muhanna would not hear of it. "There is no reason for any country or company to lie," he continued. "There is a lot of oil around."

I knew better than to ask him about Matthew Simmons. When I had met Muhanna in Washington a few weeks earlier, he had nearly broken off our conversation at the mention of Simmons's name. "He does not know anything," Muhanna said. "The only thing he has is a big mouth. Either you believe us or you don't." The truth about whether the world will have enough crude in the years ahead remains as well concealed as the millions of barrels of oil I couldn't see at Ras Tanura.

· · ·

The quandary is not Saudi Arabia's alone. If Simmons and other peak-ists are correct, the global economy can expand only so much before demand pushes against a limit in supplies again and economic growth is choked off not by toxic subprime loans but by high petroleum prices ($250 a barrel, anyone?). This scenario assumes that alternative sources of energy will not replace oil anytime soon; for now, it seems a safe assumption, alas. In America, one of the most popular nightmare scenarios of peak oil is promoted in a book entitled *The Long Emergency*, which warns that an oil shortage will trigger an economic meltdown and years of unrest, anarchy, disease and starvation. The end of the suburban lifestyle, hinged to two-car families and commutes to work, school and Walmart, will be just the first casualty.

For Simmons, scenarios of postpeak calamity were intriguing but diversionary. He preferred to focus on the question of what was happening rather than what might happen. The onetime debater found a precise topic—Ghawar, the treasure of Saudi treasures—and drilled into it (figuratively). Ghawar is the largest oil field in the world and has produced about 60 billion barrels of oil so far. The field provides more than 5 million barrels a day, which is about half of the kingdom's daily output. If Ghawar is facing problems, so is Saudi Arabia and, indeed, the world. Through his research, Simmons learned that the Saudis were using increasing amounts of water to force oil out of Ghawar—a sign of a field beyond its prime. Simmons also realized that most of Ghawar's wells are in the northern portion of the 174-mile-long reservoir. That might seem benign news—when the north runs low, the south can be tapped—but it was bad news, Simmons concluded, because the south of Ghawar is geologically more difficult to draw oil from.

"Someday (and perhaps that day will be soon), the remarkably high well flow rates at Ghawar's northern end will fade, as reservoir pressures finally plummet," he wrote. "Then, Saudi Arabian oil output will clearly have peaked. The death of this great king leaves no field of vaguely comparable stature in the line of succession. Twilight at Ghawar is fast approaching."

The Saudis and their allies did not agree. Nansen Saleri, a senior Aramco figure when Simmons began making trouble, described the Houstonian as a banker trying to masquerade as a scientist. Saleri wryly stated, "I can read two hundred papers on neurology, but you wouldn't want me to operate on your relatives." Daniel Yergin, whose consulting firm, Cambridge Energy Research Associates, earns its keep by providing advice to the oil and gas industry, offered a been-there-heard-that sigh. "This is not the first time that the world has 'run out of oil,'" Yergin wrote. "It's more like the fifth. Cycles of shortage and surplus characterize the entire history of the oil industry." At the time of these let's all-just-take-a-deep-breath lines, oil cost about $50 a barrel. Prices would nearly triple before being taken down by the global economic crisis, which was like cold water on the demand for oil.

The shots from Yergin and the Saudis delighted Simmons, who had not become wealthy by being an impeccable follower of convention. Simmons knew the risks he ran with peak oil—that people who shout "the end is nigh" do not tend to be treated well by peers or history. He noted in his book that in 1979 the *New York Times* published a story under the headline "Saudi Oil Capacity Questioned." He realized that previous Cassandras had failed to anticipate new technologies like deep-water and horizontal drilling, which found new sources of oil and raised the amounts recovered from aging reservoirs. Yergin was correct to cite the errors of earlier doomsayers, but Simmons factored all of that into his research. Technology could accomplish only so much, he concluded, raising the ante by inviting the Saudis to prove him wrong. "If they want to satisfy people, they should issue field-by-field production reports and reserve data and have it audited," he told me. "It would then take anybody less than a week to say, 'Gosh, Matt is totally wrong,' or 'Matt actually might be too optimistic.'"

Curious to see a verdict from his peers, I followed Simmons into a gathering of shipping executives that took place in a restaurant at Chelsea Piers, along the Manhattan waterfront. Barges and sailboats floated past on the Hudson River. About thirty-five besuited men sat around a rectangle of tables as the host introduced Simmons, who had a rumpled aspect—thinning hair slightly askew, coat sleeves a fraction too short. His speech skipped around in the appealing way of a brilliant, eccentric lecturer. He eventually hit his talking points, and the executives listened raptly. Simmons predicted that cheap oil would soon be a memory. The man on my right broke into a soft whistle, of the sort that meant "Holy cow"—not the sort that meant "This guy is wasting my time."

As a finale, Simmons mentioned an interview in which he'd told a skeptical reporter that the price of a barrel of oil would hit the triple digits. At the time, in 2005, a barrel fetched $50.

"A hundred dollars?" the journalist had scoffed.

Simmons issued another of his mischievous smiles as he recalled what he'd told the disbelieving journalist.

"I wasn't talking about *low* triple digits."

Whom to believe? Before visiting Saudi Arabia, I was advised by several oil experts to try to interview Sadad al-Husseini, who had just retired after serving as Aramco's top executive for exploration and production. I faxed him in Dhahran and received a quick reply; he agreed to meet me. A week later, after I arrived in Riyadh, Husseini e-mailed me, asking when I would come to Dhahran. In a follow-up phone call, he offered to pick me up at the airport. He was, it seemed, eager to talk.

It can be argued that in a nation devoted to oil, Husseini knew more about it than anyone else. Born in Syria, Husseini was raised in Saudi Arabia, where his father was a government official who acquired Saudi citizenship for his family. Husseini earned a PhD in geological sciences from Brown University in 1973 and went to work in Aramco's exploration department, rising to the highest position. Until retirement, he was a member of Aramco's board as well as its management committee. He was one of the most respected oil experts in the world.

After meeting me at the cavernous airport that serves Dhahran, he drove me in his luxury sedan to the regal yet, by Saudi standards, modestly sized villa that houses his private office. As we entered, he pointed to an armoire that displayed a dozen or so vials of black liquid. "These are samples from oil fields I discovered," he announced. Upstairs, there were even more vials, and he would have possessed more than that except, as he said, laughing, "I didn't start collecting early enough."

We spoke for several hours. His message was clear: the world was heading for an oil shortage. His warning was quite different from the calming speeches that Naimi and Yergin delivered on an almost daily basis. Husseini explained that the need to find and produce more oil was coming from two sides. Most obviously, demand was rising. Less obviously, merely to maintain their reserve base, producers needed to replace the oil they extracted. It's the geological equivalent of running to stay in place. But oil companies, Husseini said, were unable to do so. Husseini acknowledged that new fields are found from time to time—for instance, offshore near West Africa and in the Caspian basin—but he believed their output wouldn't be sufficient. With demand rising a few percentage points every year at times of economic expansion, and with the output of older and larger fields declining by a few percentage points annually, the industry needed to find vast amounts of new oil to maintain an equilibrium of supply and demand. "That's like [finding] a whole new Saudi Arabia every couple of years," he said.

He spoke patiently and firmly, like a ship's captain spotting an iceberg and explaining to his passengers that it was time to put on their life preservers. He did not disclose precise information about Saudi reserves—that could land him in jail—but he was unusually forthright and pessimistic for an insider. Traditionally, the Saudis have had an excess of production capacity that allowed them to control the market in times of emergency. In 1990, when Iraq's invasion of Kuwait shut down not only Kuwait's supply of oil but also Iraq's, the Saudis upped their output to address the shortfall. They did the same when America created jitters by invading Iraq in 2003. The Saudis functioned, as they always had, as the central bank of oil. Husseini argued that those days were over—that at times of global economic expansion, the Saudis could not satisfy the world's thirst for more.

At the energy conference in Washington, James Schlesinger, the moderator, conducted a question-and-answer session with Naimi at the conclusion of his speech, and one of the first questions involved peak oil. Might it be true, Schlesinger asked, that Saudi Arabia was nearing the geological limit of its output?

"I can assure you that we haven't peaked," Naimi replied. "If we peaked, we would not be going to 12.5 and we would not be visualizing a 15-million-barrel-per-day production capacity. . . . We can maintain 12.5 or 15 million for the next thirty to fifty years."

Husseini told me that the 12.5 million target was realistic but anything beyond it was not. Even if output can be ramped up to 15 million barrels a day, geology may not be forgiving. Fields that are worked too hard can drop off quite sharply, in terms of output, leaving behind large amounts of oil that, with better reservoir management, would have come to the surface. This is known as "trapped oil"; the haste to extract more oil can lead to less oil being extracted. "It's not sustainable," Husseini said. "If you are ramping up production so fast and jump from high to higher to highest, and you're not having enough time to do what needs to be done, to understand what needs to be done, then you can damage reservoirs."

On this, Husseini was not alone. I talked with Nawaf Obaid, a Saudi oil and security analyst regarded as being exceptionally well connected to the Saudi leadership. "You could go to fifteen, but that's when the questions of depletion rate, reservoir management and damaging the fields come into play," he said. "There is an understanding across the board within the kingdom, in the highest spheres, that if you're going to fifteen, you'll hit fifteen, but there will be considerable risks . . . of a steep decline curve that Aramco will not be able to do anything about."

Even if the Saudis are willing to try for 15 million barrels a day with the world economy recovered, Husseini pointed out a practical problem. Saudi Arabia would need to drill a lot more wells and build a lot more pipelines and processing facilities. During times of expansion, the oil industry suffers a deficit of qualified engineers and of equipment and raw materials—for example, rigs and steel. Husseini said that such things cannot be wished from thin air to meet demand. "If we had two dozen Texas A&Ms producing a thousand new engineers a year and the industrial infrastructure in the kingdom, with the drilling rigs and power plants, we would have a better chance, but you cannot put that into place overnight."

You don't need to be a Saudi geologist to grasp the peak crisis that awaits us. A fourth-grade understanding of math will do. In 2004 the Energy Information Administration (EIA), which is part of the U.S. Department of Energy, forecast that by 2020 Saudi Arabia would produce 18.2 million barrels a day, and that by 2025 it would produce 22.5 million barrels a day. Those estimates were not based on hard data about Saudi supplies but on the expected demand of the world's oil consumers. The figures simply assumed that Saudi Arabia would be able to supply whatever it was called on to supply. A year later, the EIA revised those figures downward, but not because of new and accurate information about world demand or Saudi capacities. The United States changed the figures because the original ones were so patently unrealistic.

"That's not how you would manage a national, let alone an international, economy," Husseini said. "That's the part that is scary. You draw some assumptions and then say, Okay, based on these assumptions, let's go forward and consume like hell and burn like hell." When I asked whether Saudi Arabia could pump 20 million barrels a day—about twice what it is now producing from fields that are not getting younger—Husseini paused. It wasn't clear if he was trying to figure out the answer or if he needed a moment to decide if he should utter it. He finally replied with a single, forceful word: No.

"The expectations are beyond what is achievable. This is a global problem . . . that is not going to be solved by tinkering with the Saudi industry."

Time has been good to Matt Simmons. A doubling of prices after the publication of his book prompted even Chevron and British Petroleum to announce that peak oil is a near rather than a distant event. The recession that followed, suppressing the need for oil, only put off the reckoning. We face an era of scarcity that involves higher prices for oil and fiercer competition for what's left. We are a foggy-headed boxer on his knees, unaware of the blow that awaits us.

2

Plunder

I stood beside a dirt road with some of the poorest women in Africa, waiting for one of the richest men in Africa.

Along with a few hundred unfortunate citizens of Equatorial Guinea, I was trying to avoid heatstroke at a sweltering spot where the dirt of the jungle met an unevenly paved road on the outskirts of Ebebiyin, a hungry and wary town that reminded me of a stray dog whose ribs poke hard at its skin. Around me, women swayed to a rhythm that was hard to resist, even though the lyrics were not of a can't-stop-dancing variety: "We await you Mr. President / We are happy to see you / You are the people's president." In the near distance, a cloud of Martian dust heralded the arrival of the leader whose visage was on their T-shirts.

President Teodoro Obiang's motorcade consisted of forty vehicles with enough firepower for a small war. In the lead were army-green trucks carrying elite soldiers in black ninja outfits. The jeeps in front of his armored Lexus SUV carried his Moroccan bodyguards—the president does not trust his own people to protect him—some of whom were perched on running boards, clutching machine guns aimed at the crowd. The president seemed to be invading rather than visiting. His personal armada was a projection of fear, not strength, because uneasy lies the head of a man who clutches a nation's wealth. Obiang, whose salary was reportedly $60,000 a year, had recently been discovered to control bank accounts exceeding $700 million.

Teodoro Obiang, president of Equatorial Guinea, at a parade in Ebebiyin

The presidential Lexus halted amid Ebebiyin's chickens-in-the-road squalor. Obiang strolled up the street, shaking hands with the people who lined the broken sidewalks as they shouted and gestured a bit awkwardly, unsure of the precise calibration of enthusiasm and piety desired. His posture was regal, almost rigid; he can turn on the populist charm, as dictators must do on occasion, but he wants everyone to know that he is not a common man. On this day, his strict posture and expensive suit conveyed a particular and intentional message: I am the leader, you are my subjects.

Obiang had traveled to Ebebiyin to celebrate the thirty-sixth anniversary of independence from Spain, but the weekend-long party had more to do with his alleged brilliance than the end of colonial rule. This is one of the keys to retaining absolute power anywhere: the nation and the dictator should be regarded as conjoined entities, so that the health of the former is impossible without the latter. Obiang enjoys being called "El Libertador," suggesting that he freed the country from the nightmares preceding his longed-for intervention. It is not mentioned that he "liberated" the country only from a postcolonial genocidal regime in which he was an instrumental enforcer.

The highlight of the weekend was a parade down Ebebiyin's finest stretch of asphalt. On the morning of the parade, I stopped by a two-story hospital a few hundred yards from the rutted tarmac upon which soldiers and workers would march in honor of their leader. The hospital had almost no medicine. A Cuban doctor—Equatorial Guinea does not have a medical school, and few of its citizens are licensed doctors—unlocked a storage room to show me the supplies, which consisted of bandages and a few bottles of aspirin and other pills. A place for dying rather than healing, the hospital had just received a dash of much-needed attention from the government: it had gotten a fresh coat of paint because it was visible from the reviewing grandstand and needed to look nice for the dignitaries.

There is a political price for having no medicine for the people but lots of booze for the elite (a VIP reception on the eve of the parade, to celebrate a leadership award Obiang bestowed upon himself, had featured enough Jack Daniel's to inebriate Las Vegas). A few hours before the parade, several hundred soldiers—not the presidential battalion but underfed conscripts whose jungle checkpoints served as opportunities to cajole spare change from civilians—were ordered by the Moroccan bodyguards to break down their weapons to show they had no bullets. The Moroccans supervised this task with glares that made it clear that everyone they regarded was suspect.

The parade was led by the bulletless soldiers, goose-stepping in a manner that evoked East Germany circa 1976, though with less than Teutonic precision. I sat beside the grandstand, under an umbrella for protection from the occasional rain and the painful sun that was like a laser to my head. Delegations from town after town marched by with banners saluting the president, who sat in a cushioned chair in the grandstand, paying occasional attention to the exertions below him.

Just as the festivities settled into mind-numbing redundancy, I noticed a trio of American flags coming up the road, carried by a delegation of local men and women whose banner said they were from ExxonMobil. They also carried white ExxonMobil flags and wore T-shirts imprinted with the company's logo. They were followed by delegations from Halliburton, ChevronTexaco and Marathon, all of them hoisting corporate banners, American flags and celebratory placards that hailed the wisdom of the president. They were not the only representatives of Big Oil: American executives, having flown into town on corporate helicopters, sat in the VIP grandstand, mingling with dozens of diplomats and military attachés, including a colonel from the United States Army Special Forces.

It might seem odd that dignitaries from the world's largest countries and executives from the largest companies would attend a dismal parade in a scorching corner of an oppressive nation so small it had just a few traffic lights and not a single bookstore. This was testimony to the reach of oil, which was turning the powerful into plunderers.

Just as every unhappy family is unhappy in its own way, every dysfunctional oil country is dysfunctional in its own way. A distinguishing feature of Nigeria is that it suffers all-against-all mayhem in which generals, ministers, oilmen, rebels and village chiefs are cogs in a national corruption machine. In Ecuador, oil led to the contamination

Parade in Ebebiyin, Equatorial Guinea

of a swath of the Amazon and left the government poorer, because it accumulated billions of dollars of debt it could not pay off. In Saudi Arabia, the financing of Islamic extremism is directly connected to oil. But Equatorial Guinea has been shaped in a different way. Oil consolidated power and wealth into Obiang's hands and turned American companies and America's government into his all-too-witting accomplices. If there was a silver lining to the petrodisasters of the twentieth century, it was that the same mistakes would not be made again. Equatorial Guinea became an oil exporter in the 1990s, by which time the world's politicians, bankers and oilmen had promised to do a better job.

When I visited in 2004, the results were in. Thanks to oil, Obiang enjoyed a global embrace that had enabled him to become one of Africa's richest men. The Western dignitaries at the Ebebiyin parade, though not stirring from their shaded seats, were carrying out a demonstration of their own—showing, by their satisfied presence, that they were on the side of dictatorship. Of course, it's common for Western nations to embrace dictators, but the Mugabes and Marcoses of the world tend to be abandoned if their rule becomes too odious. The calculus is different if a dictator has oil. The substance not only offers itself as a treasure to be stolen; it can become a political amulet that protects the thieves from abandonment or punishment. Obiang, who accumulated friends rather than enemies, was proof. I learned this quickly while in the country, and the steepness of my learning curve was fortunate, because within two weeks Obiang became bothered by my presence. I was accused of being a spy and expelled.

Equatorial Guinea has a population of just over 600,000 people, spread over several islands in the Gulf of Guinea and on a postage stamp of land between Gabon and Cameroon. It was one of the last African nations to become independent, and for decades its island capital, Malabo, remained a throwback to colonial times, with nineteenth-century architecture and a pace of life that recalled a sleepy Spanish town in the early Franco years. Before the oil boom, itinerant dogs could nap in the middle of streets that had more potholes than cars. Even as the oil rush was taking hold, Malabo, which had a population of perhaps 100,000 souls, retained a lost-in-time vibe. My first visit to the Ministry of Information required a jolting drive on a road that had become a floodplain after a rain shower. There was no runoff system, and there wasn't much asphalt in Malabo. When I got to the ministry, there was no electricity; it was dark and stifling and felt like the setting of a 1950s Graham Greene novel.

I wandered on foot through Malabo as much as possible, because petty crime was not a concern. Nor did anybody tug at my sleeve to ask for money or try to practice his English on me. I had the sensation of being invisible, though I was the only Caucasian on these African streets. Few people even looked me in the eye. This was unusual, and there was an explanation.

Equatorial Guinea may be the most brutalized country on earth. In 1968 the Spanish handed control of their colony to a civil servant, Francisco Macias, whom they believed to be pliant. Macias, whose self-bestowed titles included "The Sole Miracle of Equatorial Guinea," turned into a mass killer, murdering or driving into exile a third of the population. Some victims were crucified, and others shot in a soccer stadium, their screams drowned out by a military band playing the song "Those Were the Days." Macias combined the worst aspects of Idi Amin and Pol Pot but evaded global notoriety because his suffering nation was so small and of no geopolitical interest. Spain, its former colonial ruler and at the time still a dictatorship itself, paid little attention to his bloodbath.

Macias surrounded himself with relatives from his hometown of Mongomo. A trusted nephew, Teodoro Obiang, was responsible for security in the capital, and by most accounts he fulfilled his duties avidly. At Black Beach Prison, Obiang allegedly supervised the torture and murder of inmates. Obiang began fearing for his own life after one of his brothers was killed by Macias, whose insanity now threatened his confidants. Obiang made his move, staging a successful coup in 1979. Macias was sentenced to death in a show trial held at Malabo's only movie theater; the deposed madman was displayed in a cage suspended from the ceiling. Macias was executed by a firing squad composed of Obiang's newly hired bodyguards from Morocco.

Obiang inherited a moribund nation. The economy consisted of anemic exports of cocoa and coffee. Obiang didn't do much to improve matters. The kindest thing that could be said about his regime was that it was charmingly inept. One of his aides was caught smuggling marijuana into America when he left a trail of pot as he walked through John F. Kennedy Airport; his suitcase had a hole in its side. Obiang, who enjoyed playing tennis in a nation where ownership of a soccer ball was a sign of wealth, was mentioned in the foreign media when lists were published of the world's worst dictators; he usually made the top ten.

His status was deserved. An opposition leader announced one day that Obiang "has just devoured a police commissioner. I say devoured, as this commissioner was buried without his testicles and brain." Cannibalism has been denied by the government, but what's undeniable is that Obiang did not treat his rivals with kindness. John Bennett, a former U.S. ambassador to Equatorial Guinea, told me of an opposition politician who returned from exile, unaware that Obiang remained displeased. Police thugs went to the politician's hotel room, threw him to the ground and dragged him away by his feet, his head bouncing off a set of marble stairs. Later, he was beaten to death with staves; his demise was officially listed as a suicide. Bennett didn't last long, either. He was evacuated in 1994 after receiving a death threat from an Obiang aide who soon became foreign minister. The American embassy was closed the following year. When we spoke before my trip to Equatorial Guinea, Bennett described it as "the world's finest example of a country privatized by a kleptomaniac without a scintilla of social consciousness."

At almost the same time Bennett had to leave, several hundred million barrels of oil were discovered in the country's waters in the Gulf of Guinea. That's a minor amount in the grand scheme of things—Nigeria has more than 35 billion barrels of oil, Saudi Arabia more than 260 billion—but it's a destiny-altering amount for a nation whose population is smaller than that of Memphis. The way was led in the early 1990s by a small American company, Walter International, which then sold its valuable concession to a bigger firm that had the know-how and financing to build an extraction infrastructure of wells and pipelines. Once the first fields were opened, larger companies, such as Marathon, bought rights to them and found yet more fields. This is often the way things work in the oil industry—there is a food chain of extraction rights and discoveries.

Although there are variations, usually a company pays a government a nonrefundable fee for the right to search for oil or gas. Depending on the likelihood and difficulty of success, the fee can range from a few million dollars to tens of millions or more. The poorer and less sophisticated a country is, the less it tends to receive, because its bargaining position and financial prowess are not terribly strong. Corruption plays a role, too. Quite famously, Standard Oil of California paid 55,000 pounds in gold for the right to search for oil in eastern Saudi Arabia in the early 1930s. At the time, the Saudi king was in desperate need of cash because his new country was deeply in debt and on the verge of falling apart. At the outset, the American companies that now extract oil from Equatorial Guinea reaped unusually large profits—the result, according to the International Monetary Fund, of Obiang and his financial advisers not understanding or not caring to understand how much they were giving away. This is one of the ways a country's wealth is stolen. Corrupt dictators and officials, after taking bribes or negotiating lucrative side deals for themselves, award exploration and production contracts that are unduly generous to the companies on the other side of the bargaining table. The consequence is that crooked officials become rich but their country receives less for its resources than it should.

With oil-prospecting rights in hand, a company spends what can be considerable amounts on geological surveys and exploratory wells. The risks are high—if oil or gas is not found, the exploration costs are lost, usually tens of millions if not hundreds of millions of dollars, on top of whatever was paid for the mere right to explore. The gamble is offset by potential profits, because a company that finds oil or gas often has the contractual right to a share of revenues gained from extracting the treasure; at the least, it has an inside track on the next round of bidding. Walter International got lucky—it found oil under Equatorial

Guinea's waters. The small company was soon bought out by a larger firm. Big Oil had found, as they say in the industry, a new "play."

The "majors," as the largest companies are known, poured money into the country—more than $7 billion by 2008—to build offshore platforms and onshore facilities to extract nearly 400,000 barrels of oil a day and considerable volumes of natural gas. Just a few miles from Malabo, Marathon built a $1.5 billion plant to turn natural gas into a frozen liquid that could be shipped to America and Europe. Malabo's landscape was altered. At night, if you stood on the patio of the Hotel Bahia, where in the 1970s Frederick Forsyth wrote *The Dogs of War*, and where the political elite continued to gather for drinks when I was there, you could see the flares of Marathon's plant, which was about ten miles away. In fact, you could not miss the flares—their light was strong enough to cast shadows at the Bahia.

In a geological blink of an eye, Equatorial Guinea became the third-largest energy exporter in sub-Saharan Africa, after Nigeria and Angola. Malabo boasted nonstop flights to Texas; they were known as the "Houston Express" and were filled with oilmen. By the time of my visit, gross domestic product had soared an almost unimaginable forty-fold from the pre-oil days. The country was exporting more oil and gas per capita than Saudi Arabia. If the revenues were spread evenly around the country, the people in Equatorial Guinea would be among the richest in the world.

The reality is fantastic only in the worst of ways.

Even if Mother Teresa were president of Equatorial Guinea, the odds would be stacked against her subjects getting rich from the country's mineral wealth. That's because it is not just the thieving activities of government officials that make it hard for average citizens to benefit from oil booms. The globalization of labor, combined with the small number of workers needed for capital-intensive oil projects, ensured that most Equatorial Guineans would watch others profit from the boom.

Let's begin at Marathon's natural gas facility, the one whose immense flares brightened (and polluted) the night sky. Little of the

$1.5 billion Marathon and its minority partners spent on the facility entered the local economy because the plant was built by thousands of foreign workers who lived on the construction site and sent their paychecks home to Manila and Houston. Even for manual labor—digging ditches and the like—workers were flown in from India and Sri Lanka. I wouldn't have understood the economic logic had I not been driven around the plant by its genial manager, Rich Paces, a Texan, in his SUV.

As we moved around the site I noticed that almost all the workers were South Asian, while managers were American or European. I knew that Equatorial Guinea's labor force included few college-educated managers, but there was no shortage of young men who could pound nails. Why weren't they working here? Paces explained that the Indians and Filipinos had previous experience on large projects of this sort, so they required little instruction. They knew how to use welding torches, they knew how to avoid injury from the heavy machines on the site, and they could be counted on to work twelve-hour shifts without complaint. Also, they were not hobbled by malaria or yellow fever, which were rife among the native population. The few locals working at the plant were hired under a quota written into the company's contract with the government.

"We're under contract to hire them, and it's the right thing to do," Paces said. "But if we didn't have those limitations, we'd entirely staff it up with low-cost Filipinos and Indians."

The plant—like many oil installations in the developing world—could have been on the moon for all the benefit it offered local businesses. Thanks to just-in-time supply networks that span the globe, Marathon saved money by importing what it needed rather than working with unfamiliar local suppliers. Instead of buying cement from a Malabo company that might not deliver on time, Marathon built a small cement factory on the construction site. Raw materials were imported, and the factory would be dismantled when construction ended. The trailers in which the Asians lived were prefab units—no local materials or local labor had been used to build them. The plant had its own satellite phone network, which was connected to the company's Texas network—if you picked up a phone you would be in the Houston area code, and dialing a number in Malabo would be an international call. The facility also had its own power plant and water-purification and sewage system. It existed off the local grid.

"Almost everything has to be imported," Paces explained.

How about paint? I asked.

"Imported, sure," he replied.

Portable toilets?

"Yes."

Equatorial Guinea had a lumber industry, so I asked whether the wood, at least, was local.

"No, imported."

Food?

"Most of it gets imported."

Construction cranes?

"You have to be self-sufficient out here."

I pointed to the small rocks that had been lined up to denote the shoulders of a dirt road on the site.

"Those are local rocks, but importing them would be cheaper," he said.

That night, I strolled through the center of Malabo and found a few hubs of local job creation. Their names were La Bamba and Shangri-La, and they were open-air bars that played country music to attract the roughnecks from Texas and Oklahoma. The teenage girls who swarmed these establishments were dressed in skirts made for whistling at. The oilmen called them "night fighters" because they battled each other for the chance to spend an evening with one of the rich foreigners. As I walked past La Bamba, several girls trotted out of the bar to chat me up. The men in Malabo might not find jobs in the oil industry, but it was clearly possible for their desperate sisters to earn a few dollars.

The oil companies provided few jobs to local people, but they were paying royalties—at least in the hundreds of millions of dollars a year—for the privilege of extracting and selling Equatorial Guinea's hydrocarbons. Where was *that* money going?

It was no secret that Obiang lived far beyond the means of his official salary. He belonged to the class of dictators who do not feel the need to obscure the comfort they grant themselves. Conspicuous consumption is a manifestation of greed as well as a way to project power. In a country like Equatorial Guinea, anybody could kill you with a gun, but how many people could afford to live in a mansion with *two* tennis courts? Obiang had built, along the well-traveled road from the airport into Malabo, several mansions for himself and his wives. Although the dictator was not always in evidence—the airport road was closed to other vehicles when his motorcade traveled on it—his wealth was well advertised. He bought, for $55 million, a Boeing 737 in which the bathroom fixtures were gold plated. When the plane was flown into Malabo's airport, Obiang was on hand to inspect it and boasted to reporters, "This plane elevates the image of our country in the developing world." His indulgences were almost modest compared with those of his eldest son, Teodorin, who bought luxury homes in at least two countries, including a $35 million estate in Malibu. To round out the portrait of playboy extraordinaire, young Teodorin dated the rapper Eve and invested in a hip-hop label, TNO Records. For a weeklong Christmas cruise, Teodorin rented, for a reported $700,000, a monster yacht that belonged to Microsoft cofounder Paul Allen. It should be noted that Teodorin's official salary at the time, as the minister of forestry, was $48,000 a year.

The first family could not spend all the money coming in from the oil companies, hundreds of millions of dollars a year. Yet little of this bonanza was used for improving the lot of ordinary people. When I visited the minister of education, I noticed that he worked out of an office that had a bare light bulb for illumination and a couch the Salvation Army would have rejected. The most valuable item in his office was his Rolex, which suggested that his needs received more attention than his ministry's. If he required more light, it seemed he would need to buy a bulb himself or swipe one from a subordinate. It was like that in every government building I visited, including the few schools that existed at the time; there was dilapidation at every stop. More than a decade after oil's discovery, nearly half of the children under the age of five remained malnourished, few households had running water or electricity, and soldiers at checkpoints continued to demand bribes— *"cerveza"* they would say to their victims (including me). There was little choice but to hand over beer money and hope for better times. "The staggering increases on paper stand for little," a British report noted. "Equatorial Guinea's wealth is concentrated in the hands of a tiny elite, so oil revenues do not benefit the majority and do not stimulate the economy as a whole."

The money trail is no mystery. Old ladies and petty criminals can hide their savings in mattresses, but oil dictators cannot do the same with hundreds of millions of dollars. Stealing oil revenues is not like holding up a grocery store. You need to get the oil and gas out of the ground, you need to ship and sell it to foreign markets, you need to put the revenues into bank accounts, and you need to find ways to share some of this money with the family and friends who are the political elite that help run your country and keep you in power. You need, in other words, a lot of help. Usually this can be done without indictment; offshore banking is made for such tasks. But Obiang made a mistake that allowed the rest of the world to understand the dynamics of making a nation's oil wealth disappear. Instead of stashing the millions in a numbered Swiss account, he selected a niche bank in Washington, D.C., that was willing to abet his plan but was incompetent at doing so.

A ritual indulged in by men of great wealth is the luncheon with their bankers. The wealthy client is ushered through the bank's private entrance and is conveyed in a special elevator to a wood-paneled dining room, where he is flattered and fed like a king. If the client makes a joke, the laughter is fulsome. If he suggests a certain investment priority, his opinion is received as brilliant. This, more or less, was the white-glove treatment granted to Obiang by Riggs Bank, which for more than a century had offered its discreet services to the powerful and well connected in Washington, D.C. When Obiang visited the

United States in 2001, a lunch was held by the bank, where Obiang had become, thanks to enormous deposits into his accounts from Exxon and other oil firms, the largest client. His fawning hosts—the bank's president and other executives—sent him a grateful follow-up letter, which congressional investigators later obtained.

"We hope this letter finds you well and rested after your last busy trip to Washington," their missive began. "We would like to thank you for the opportunity you granted to us in hosting a luncheon in your honor here at Riggs Bank." The letter expressed their "gratitude" for the chance to serve Obiang and promised to "reinforce your reputation for prudent leadership and administration as you lead Equatorial Guinea into a successful future."

Obiang started his relationship with Riggs in 1995, when he decided that it would be easiest to use an American bank as a parking spot for the money he was beginning to receive from American oil companies. After all, why go to the trouble of using a shady Swiss outfit when a respectable American establishment was willing to do the job? Obiang assumed that his solicitous bankers at Riggs could handle his affairs in a way that would defer inconvenient questions from the U.S. Treasury Department. Obiang and his family and his associates

Riggs Bank in Washington, D.C.

opened sixty accounts at Riggs, some for official purposes, others for personal affairs. Deposits in those accounts reached $700 million. In tandem, Obiang and companies he owned opened accounts in countries with bank secrecy laws—the kinds of places where money goes to disappear for good—and filled those accounts with funds that came from Riggs, which acted like a traffic dispatcher in forwarding its clients' money.

The primary Riggs account, known as the "oil account," was where energy companies deposited royalty payments, and this account often held tens of millions of dollars at a time. It was used as a holding place for the money before it was transferred into other accounts at Riggs and elsewhere. There is no suggestion that payments into this account were bribes—the funds were owed to Equatorial Guinea by the oil companies—but the companies showed no interest in knowing what happened to this money. They did not seem to care that their payments were being handled in a fashion that screamed theft: the oil account at Riggs was controlled not by Equatorial Guinea's government but by its president. (Imagine a company making its tax payment not to an account controlled by the IRS but to an account controlled by Barack Obama.) Obiang's regime wired—without objection or scrutiny from Riggs—$35 million from the oil account to mysterious companies in banking havens with strict secrecy laws.

Then there were the "investment accounts" of Obiang's regime. In 2003, the value of these accounts fluctuated between $300 million and $500 million. In effect, these were Equatorial Guinea's financial reserves—the money that was gained from oil sales. It is unusual for the bulk of a country's financial reserves to be held in a private foreign bank, especially a relatively minor one like Riggs. And it is even more unusual for transfers or withdrawals from these accounts to require just the president's signature. Obiang actually did relatively little with them, aside from occasional transfers to the accounts in banking havens. Instead of being invested in schools or hospitals or light bulbs for the minister of education's office, and while his compatriots died of malaria and hunger-related diseases, most of the money stayed at Riggs.

On occasion, the situation resembled a Charlie Chaplin movie. The Riggs banker who oversaw the accounts, Simon Kareri, twice went to the Equatorial Guinean embassy, a mile from his office, to pick up suitcases that weighed sixty pounds and contained $3 million in plastic-wrapped stacks of hundred-dollar bills. He dragged them back to Riggs and deposited their contents unquestioningly into one of Obiang's accounts. ("Sir," Kareri wrote in a memo to his superior, "I wish in due course you will get to know the President of Equatorial Guinea and witness his simplicity first hand.") The bank also received cash deposits of more than $1.4 million into accounts belonging to one of Obiang's wives. In those cases—as with other large cash deposits— Riggs did not file timely or accurate "suspicious activity reports" to regulatory authorities, as required whenever a bank suspects or should suspect that a transaction might involve illicit activities. Riggs was caught up in an age-old fever. In an e-mail, a senior Riggs banker excitedly predicted more deposits from Equatorial Guinea and explained, "Where is this money coming from? Oil—black gold—Texas tea!" It was such a free-for-all that Kareri transferred more than $1 million into an offshore company he controlled. It appears the dictator who was siphoning money from his country was himself being pickpocketed by his banker.

It all came crashing to the ground when Senate investigators, reacting to stories about Obiang and Riggs by investigative journalist Ken Silverstein, launched their probe. In 2004 the Senate released a report entitled "Money Laundering and Foreign Corruption: Enforcement and Effectiveness of the Patriot Act, Case Study Involving Riggs Bank." A sexier but no less accurate title might have been "How a Despot Looted His Country with the Help of American Bankers and Oilmen."

American oil companies were involved in far more than ordinary royalty payments to Obiang's regime. They were also making a variety of payments that seemed geared toward rewarding important figures in Equatorial Guinea. The Senate report got straight to this point: "Oil companies operating in Equatorial Guinea may have contributed to corrupt practices in that country by making substantial payments to, or entering into business ventures with, individual E.G. officials, their family members, or entities they control, with minimal public disclosure of their actions." Among the payments were more than $4 million that various firms, including Exxon, Chevron, Marathon and Hess, provided for tuition and living expenses of Equatorial Guinean students abroad. These are the sorts of scholarship programs that companies like to portray as shining examples of their enlightened and generous charity. But according to the report, most of the students appeared to be "children or relatives of wealthy or powerful E.G. officials." These are akin to bribes in kind; instead of slipping $50,000 to a government official, a company pays the college expenses of the official's son.

Between 1995 and 2004, millions of dollars from these oil firms were deposited into Riggs accounts for what appeared to be real estate or business deals. Payments were made to, among others, the president's wife, the interior and agricultural ministers, and at least one well-placed general. For several years Exxon paid between $135,000 and $175,000 to Obiang's first wife, Constancia Nsue, to rent a compound that houses its workers and offices. (I asked to visit the compound while I was in the country, but Exxon officials refused my requests. I showed up at the compound one day but was turned away by security guards. I noticed, behind them, an expanse of well-tended lawns, finely paved roads and neatly ordered cottages—a contrast to the shacks most local people live in.) Exxon also paid $236,160 to a firm owned by the interior minister; it is unclear what the payments were for. The prize for the most unusual deal went to Amerada Hess Corporation, which rented property for $445,800 from a fourteen-year-old relative of Obiang's.

The oil companies say these contracts were not bribes but, rather, payments for necessary goods or services provided by people who just happened to be the president's relatives and ministers. And it has to be said that if the oil companies were indeed looking to purchase goods or services, other than from friends in important places, they may not have had much choice. The Senate report noted that in Equatorial

Guinea, as in most kleptocracies, the ruling family, the government and the business elite are one and the same. If an oil company needed to hire security guards for a warehouse in Malabo, there was only one local company licensed to provide guards: it was called Sonavi, and it was owned by Armengol Nguema, the president's brother and the director of national security. "How oil companies can and should respond to this situation raises a number of difficult policy issues," the Senate report acknowledged.

There is little doubt that the local firms American oilmen dealt with were fronts for enriching the elite. The president's playboy son Teodorin admitted in an affidavit that he and other ministers routinely collected kickbacks on government contracts. The affidavit was filed in response to a lawsuit that sought to deprive Teodorin of two houses he had purchased in South Africa for $7 million. Teodorin needed to prove that the money with which he'd bought the houses was his own rather than the state's. He noted that he was a minister in his father's government, and that "cabinet ministers and public servants in Equatorial Guinea are by law allowed to own companies that, in consortium with a foreign company, can bid for government contracts." The affidavit went on to explain that once the contract is awarded, the "cabinet minister ends up with a sizable part of the contract price in his bank account." This affidavit was almost surreal—an honest account of dishonesty.

These were equal-opportunity rip-offs because the companies that funneled back-channel money to Obiang's regime were, in return, receiving special benefits from the regime. This takes us back to the fact that oil companies tend to get better terms from lesser-developed governments. According to an International Monetary Fund assessment, oil companies in Equatorial Guinea received "by far the most generous tax and profit-sharing provisions in the region." The government received only 15 to 40 percent of the revenues from the sale of its oil and gas; the rest went to the companies that extracted the resources. The norm in sub-Saharan Africa was for host governments to receive 45 to 90 percent of the revenues from oil and gas sales. And often, in Equatorial Guinea, the companies failed to pay the government what

they owed: the IMF said that oil and gas companies underpaid the government by $88 million between 1996 and 2001. It would seem that
the companies and the Obiang family had a wink-wink understanding:
You give us extra money here and we'll give you extra money there.
The only losers in this understanding, of course, were the destitute
people of Equatorial Guinea.

To see, on a transactional level, how plunder is achieved, and how an
oil firm might funnel a bribe to a president, it's useful to look into the
affairs of Abayak S.A., a mysterious company that was widely believed
to be owned by Obiang and that even a Riggs memo described as "a
significant earner of income for the President." Was this the conduit
through which American companies slipped money into Obiang's
already bulging pockets? The Senate report found a number of unusual
payments to Abayak. For example, Marathon negotiated a deal to purchase land from Abayak for more than $2 million; a partial payment
was executed with a check for $611,000 made out to Obiang himself.
Marathon was involved in a joint venture to operate two natural gas
plants with GEOGAM, an obscure company in which Abayak held a
75 percent stake. And so on.

But nobody seemed to know what Abayak did. Were these payments for actual goods and services?

I thought the answer might be found in Bata, the largest city on the
mainland portion of Equatorial Guinea. Like Malabo, Bata retained a
sense of postcolonial languor, with ramshackle Spanish architecture of
the one-story variety and semiderelict streets with potholes a mule
could disappear into. The pool at the town's finest hotel, which was not
fine, was filled with a foot of greenish murk, and even the hotel's
name—Pan-Africa—was a throwback to another era. Yet the biggest
office building in the country at the time, reaching to seven stories, had
just been completed there, and because it was known as Abayak's headquarters—the company's name was emblazoned at the top—I visited it
in the hope of talking with an executive or two.

At the ground-floor reception area, I was told that the firm's offices were on the top floor. When I went there, I found that four of the six offices were vacant and unfurnished. Doors to the two remaining offices were locked and unmarked. If this was Abayak's headquarters, it seemed unfathomably modest for a firm that had been selected as a strategic partner by the largest oil companies in the world.

Perhaps the receptionist was wrong; maybe Abayak's offices were on another floor. I checked every floor and saw that the offices were either empty—most were—or occupied by other entities. Even the Ministry of Information official who accompanied me was flummoxed. Where was Abayak? And, more to the point, *what* was Abayak?

There were answers in Malabo. A British businessman told me that as far as he knew, Abayak functioned as a vehicle through which payments were made in exchange for the president's approval of business projects. An African banker I talked with called Abayak a "holding company" that, he confirmed, had no offices. Indeed, there was no Abayak office in Malabo that I could locate. It was not possible to ask the president to solve the mystery—my requests for an interview were declined—so I went to the next-best source, his son Gabriel Nguema Lima, who, in the fashion of family regimes, was vice minister of mines, industry and energy. Nguema was in his twenties.

His office was in the ministry headquarters, a modest two-story building where a rooster was pecking around the front yard. His office had a flat-screen computer but was not large; in most governments it would house a midlevel civil servant. It was air-conditioned by a wheezing wall unit that belonged in a junkyard. Adorning the wall was Nguema's diploma from Alma College in Michigan and his varsity soccer letter from a Michigan prep school, Cranbrook Kingswood. Odd as it seemed, this presidential son was a pseudopreppy, and odder still, given the ways of his tennis-playing father and his Malibu-dwelling half brother, he didn't seem ostentatious. He was, nearly everyone said, the best hope for the country's future. Once his father passed from the scene, and if Teodorin could be kept from power, perhaps Nguema could take charge.

Nguema initially described Abayak as a business with operations in the cement and cocoa sectors. His pretense did not hold up for very long. I told him I had tried to locate its headquarters and had found nothing in Malabo and had come up empty at the building that bore its name in Bata.

The president's son scratched his head.

"Um, headquarters of Abayak, that's a good question," he said, pausing uncomfortably. "I don't think they have a headquarters. The headquarters would be"—he paused again and looked at his feet—"maybe my father's house."

How did American companies defend their dealings with Obiang? When its report was published, the Senate held a hearing at which the head of Riggs testified, as did senior executives of Exxon, Marathon and Hess. Oil companies tend to be quite adept at avoiding uncomfortable questions—they routinely refuse to make their executives available for interviews on controversial topics—but a summons from Congress cannot be refused.

Lawrence Hebert, president and chief executive of Riggs, expressed regret that his bank did not "fully meet the expectations of our regulators." He blamed the absence of suspicious activity reports—which are supposed to be submitted to regulators when large cash payments are made—on a subpar computer system.

Senator Carl Levin was amazed.

"Mr. Hebert," he said, "you don't need a computer system to realize suspicious activity when you've got sixty pounds of cash there being walked into the door with a suitcase."

Senator Levin was just warming up. He noted the effusive letter Hebert wrote to Obiang after their lunch in 2001.

"How do you write that stuff to a man as abominable as this guy?" Levin asked. "How do you basically live with yourself?" Levin also said, "After this information became ever more public, you continued to do business with him."

"Well, we watched him closely," Hebert replied. "We took prudent steps to be very careful with this gentleman."

"Who you calling a gentleman?" Levin shot back. "Let's call him a dictator."

Later came the oil executives. Andrew Swiger, then an executive vice president at ExxonMobil, was the first to testify.

"The business arrangements we've entered into have been entirely commercial," Swiger said. "They are a function of completing the work that we are there to do, which is to develop the country's petroleum resources and, through that and our work in the community, make Equatorial Guinea a better place."

"Make it what?" Levin asked.

"A better place," Swiger replied.

Levin concluded, "I know you're all in a competitive business. But I've got to tell you, I don't see any fundamental difference between dealing with an Obiang and dealing with a Saddam Hussein."

It will not come as a revelation that with the exception of Senator Levin and a few other voices in Congress, the U.S. government was as pliable a friend to Obiang as Riggs and Exxon. Grand theft—*national* theft—cannot be accomplished without the involvement of a lot of institutions, including foreign governments. The arc of U.S. relations closely followed Equatorial Guinea's oil production: the more Equatorial Guinea exported, the more that was taken from its people, the better its relations with Washington. It did not matter that Obiang was a dictator and that his foreign minister had threatened to kill the American ambassador in the 1990s and that the American embassy had been closed after that. One of oil's darkly magical properties is that it erases inconvenient memories.

In February 2001, Obiang was guest of honor at a private Washington luncheon organized by the Corporate Council on Africa, a lobbying group composed of American companies with investments in Africa. The council hailed Obiang, in a bio handed out at the luncheon, as the "first democratically elected president" of Equatorial Guinea. (In his most recent election, which even the State Department had described as flawed, Obiang won 97 percent of the votes.) According to

Ken Silverstein, the luncheon was held at the Army and Navy Club, and Obiang was seated at the head table with fawning oil executives and senior State Department officials, including Assistant Secretary of State for African Affairs Walter Kansteiner. After hearing speeches that praised his country as "fabulous" and "the Kuwait of Africa," Obiang returned the kindness by declaring that "we can promise American companies that their investments are guaranteed."

Soon after, the Bush administration decided to reopen the embassy in Malabo. Though Obiang's regime was no less odious than before, Secretary of State Colin Powell and Energy Secretary Spencer Abraham each met with Obiang in 2004. Those sessions were private, but in 2006, as oil prices climbed higher and West Africa emerged as a crucial supplier, Obiang met with Secretary of State Condoleezza Rice, who described him, during a press conference, as "a good friend." The Obiang family was delighted with its new relationship with Washington. As Gabriel Nguema, the president's preppy son, told me, "The United States, like China, is careful not to get into internal issues."

I don't think a novelist could have invented the next chapter. At his confirmation hearing in 2006, Donald C. Johnson, the newly nominated U.S. ambassador to Equatorial Guinea, admitted an unusual fact about the new embassy in Malabo: the building was being leased from the minister of national security, Manuel Nguema Mba. Nguema is the uncle of President Obiang, which meant the U.S. government, like Exxon and Marathon, was putting money into the pocket of the presidential family. Yet the most appalling twist was that the State Department and the United Nations Commission on Human Rights had documented a case in which Nguema Mba had supervised the torture of a political opponent. An alleged torturer was receiving $17,500 in monthly rent from U.S. taxpayers.

In Malabo, it was easy to notice one of the reasons the American government was a good friend to the Obiang regime.

I stayed in the Dynasty Hotel, which had perhaps twenty closetlike rooms and would qualify in most cities as dingy, though in Malabo it was one of the best. Because Equatorial Guinea had just begun to join the international grid, it was not yet possible to pay my bill with a credit card. The hotel, like the few others in town, accepted only cash or transfers into an overseas bank account. I chose the latter option and was told, at the reception desk, that the account was located in Shanghai. The hotel was called "Dynasty" because it was Chinese-owned. The manager, the receptionists, the waitresses in the small café, the cooks—they were all Chinese. The lamps in my small room, the desk, the bed, the phone, the armoire—all from China.

More than a century ago, it was the British competing against the Russians for mineral resources in central Asia. After that, it was Americans against Soviets in the twentieth century. Now the great race pits China against everyone else. Or so it seems. This is the modern narrative of resource competition, and it has a xenophobic tone, as though China, a newcomer, has less of a right to consume the world's resources than the rest of us. But consume they do.

When I drove from Bata to Ebebiyin to watch Obiang's parade, the road was being paved by Chinese workers. Other crews of Chinese were building a road from Bata to Mongomo. The Chinese government now imports more than $2.5 billion in oil from Equatorial Guinea every year, and China National Offshore Oil Corporation has signed a contract to explore for oil in the waters of Equatorial Guinea. This was an inevitable consequence of China's effort to acquire the natural resources it needed to fuel its economy. Most famously, China has become a key economic partner in Sudan's genocidal regime; Chinese companies are extracting oil from fields in the south of Sudan. In Nigeria, China agreed to build an eighteen-hundred-mile railway. A Chinese company bid $2.4 billion to extract oil from Angolan waters. And China is going straight to the mother lode, buying billions of dollars of oil from Saudi Arabia while working with Saudi companies to construct refining facilities in China itself.

Scary stuff, it might seem, but the fact remains that American and European companies continue to have a far larger presence in Africa and the Middle East. There were far more Texans in Malabo than Chinese. In some ways the Chinese threat was useful to American interests. American oilmen and diplomats contend that if they pull out of unsavory countries like Equatorial Guinea, the Chinese will fill their still-warm seats and apply less pressure for political change. But in Equatorial Guinea, there has been no substantive political change since American companies began investing their billions there. Much of their operations—and all of their negotiations—are conducted in conditions of near secrecy. Marathon allowed me to visit its natural gas plant outside Malabo, but Exxon, the largest foreign company in the country, refused to let me talk to any of its local executives or visit its headquarters; its gated compound was as much of a forbidden city as the real one in Beijing. If any of the American executives had been allowed to talk to me, they probably would have put forward the argument I heard from an American oilman in Quito, Ecuador, when I asked about the environmental movement's opposition to his company's plans for further extraction in that country. "The fear I have," the oilman said, "is that all the [American] companies are going to leave. Who's going to come in? The Chinese will behave as they do in China. Saying no to us is going to turn into saying yes to the wrong people."

The argument made sense in theory—America is a democracy and China is not—but in the real world of deeds and misdeeds, it can be hard to see a great difference. The American government was paying rent for its embassy to a minister accused of torture. Obiang's family was reaping huge profits from sweetheart deals with American companies. There seemed to be more in the way of reward for bad behavior than pressure to change it.

My expulsion was inevitable. I had let the authorities know I was working on a book about oil, and after I'd been in the country for ten days the minister of information, Alfonso Nsue Mokuy, began calling and texting me, demanding that we meet immediately. I sensed what was coming, because several journalists who had visited Equatorial Guinea in recent years had been expelled in a matter of days; Equatorial Guinea was, according to the New York–based Committee to Protect Journalists, one of the most censored countries in the world. (Only

North Korea, Burma and Turkmenistan were worse.) I suggested that we meet at the patio café of the Hotel Bahia, a very public place where no harm could occur.

The minister was waiting for me when I arrived, along with a man who wouldn't disclose his name but said he was an adviser to President Obiang. They did not bother to shake my hand. They were not happy.

"Peter, you have caused us enormous problems," the minister said. "The president has called me three times, and him"—he nodded to the presidential aide—"four times."

The hotel pool, just a few yards away from us, had no water, the sky was overcast and the waiters, normally attentive, kept their distance. Nothing boded well. The minister was sweating profusely, because his job and perhaps his life were on the line. In dictatorships, information ministers are expected to control the activities of journalists, and Minister Mokuy had failed to do so in my case. He told me Obiang was upset because I'd met Spain's ambassador—Spain was then refusing Obiang's request to extradite a political rival—and because he believed I was involved with a wire-service story criticizing his rule. (I had no knowledge of the story.)

My explanations failed to persuade. The minister gave me fifteen minutes to pack my bags. Then the adviser drove me to the airport, where I waited for the next flight out. A final bout of unpleasantness occurred when the minister arrived at the airport and accused me of being a spy. He searched my bags, confiscated several disks (fortunately I had backups) and threatened to take me downtown for a real interrogation. He backed off when I said President Obiang would never visit Washington again if his regime imprisoned an American journalist. I had no idea whether this was true, but the minister reacted as though I had said a magic word.

Obiang could jail and torture his subjects as much as he wished, and appropriate the country's resources without challenge, but Americans would be spared his rough treatment. This is the practice of bullies; they assault the weak and cower from the strong. America's relationship to Obiang had saved me from further discomfort—oil had saved me, to be precise—but America's influence and oil itself were of little benefit to the people of Equatorial Guinea. As the turboprop taking me to Cameroon rose into an evening storm above the airport, the only light I could see through the clouds was Marathon's flare. My last glimpse of Equatorial Guinea was fitting. Darkness and flames. A landscape of plunder.

3

Rot

Who owns it?

The discovery of oil usually unearths this question, and the answer is not simple. Is the oil owned by the farmer who works the land that sits atop the oil? The surrounding community? The state in which the community is located? The federal government in a capital hundreds or thousands of miles away? The foreign company that invested millions of dollars to find it?

Even in Norway and Canada, countries with cohesive political institutions, these questions required considerable time and effort to settle. The task is harder for countries without national identities. Not just harder but sometimes lethal, because power rather than justice can prevail in such disputes. As the oilman J. Paul Getty noted, "The meek shall inherit the Earth, but not the mineral rights." Communities in the Niger Delta, where most of Nigeria's oil was found, received little more than token payments after significant extraction got under way in the 1960s, and this accelerated a process of national breakdown. At first, there were peaceful protests, which were met with state repression. Militias were then formed to do battle with soldiers who attacked disgruntled villages. The militias and their supporters took matters into their own hands; oil workers were held for ransom, pipelines were tapped into. The militias also fought one another, because struggles for justice can develop into grabs for cash, and some militias were little more than gangs that Nigerians called "cults." Foreign companies fed

Nigerian worker at an oil spill in the Niger Delta

the conflict by providing funds to both sides: the military was paid to protect wells; the militias were paid not to attack them. The combatants were incentivized for combat. I visited Nigeria to learn how oil had turned a once healthy country, and the people who lived there, into a specimen of rot.

Midmorning wrapped a humid embrace around Port Harcourt, which is the heart of Nigeria's oil industry and has a population of about 1.5 million. Located at the mouth of the oil-rich Niger Delta, Port Harcourt is a typical Nigerian city—it is sprawling, chaotic and violent. (Guns reportedly outnumber computers by four to one.) On this morning, Dokubou Asari, a warlord, stayed in bed at the small hotel that was his temporary headquarters in the city. When awake, he was the vigorous leader of an uprising against Nigeria's armed forces and the oil companies they protected. His uprising depended, for its rallying cry and financial sustenance, on attacking some oil facilities while siphoning crude from pipelines operated by Royal Dutch/Shell, Chevron and other firms. Depending on your view, Asari was a thief, stealing from the companies a resource that was not his, or Asari was a Robin Hood, restoring to his people what was being stolen by foreign companies and a corrupt government. We would talk once he awoke.

When oil was discovered in Nigeria by geologists working for Shell, the country had a growing industrial sector and a healthy farm economy. With its British-educated elite, Nigeria's prospects were bright in 1960, when it became independent; its people were led to believe that the just-discovered treasure in the delta guaranteed a brilliant future. One of them was Annkio Briggs, a senior aide to Asari, who told me of Shell managers visiting her village with a movie projector. The villagers, most of whom had never seen a movie, gathered to watch a company film about the prosperity oil would bring. "They showed pictures of how white people lived in suburbs," Briggs recalled. "Water came out of the taps. Children were getting into cars. Like them, we would live the good life." Shell was welcomed into the Niger Delta.

Now the world's eighth-largest exporter of oil, Nigeria earned more than $400 billion from oil in recent decades, yet nine out of ten citizens live on less than $2 a day and one out of five children dies before his fifth birthday. Its per capita GDP is one-fifth of South Africa's. Even Senegal, which exports fish and nuts, has a larger per capita income. Nigeria's wealth did not vanish, as in a magic trick. It has been stolen by presidents, generals, executives, middlemen, accountants, bureaucrats, policemen and anyone else with access to it. This is what can happen in a country with weakly enforced laws and a weak sense of national identity: it becomes every region for itself, every tribe for itself, every family for itself. But the fruits of thievery are lopsided. The World Bank estimates that 80 percent of Nigeria's oil wealth has gone to 1 percent of the population. A few years ago the national police chief was convicted of stealing $98 million, and the punch line was his sentence: six months in jail—one month for every $16 million. As for the money that wasn't stolen, much was squandered on projects like the multibillion-dollar Ajaokuta steel complex, which has not made a single slab of steel.

The cruelest joke is that even if oil money is not stolen or wasted, it can nonetheless have negative economic consequences. The problem begins with the influx of foreign currency from oil sales, which seems like a stroke of great luck. When large amounts of foreign currency flood into an exporter's economy, the local currency tends to appreciate. When this happens, foreign products become cheaper to buy with the strengthened local currency while domestic products become more expensive for foreigners to buy. As a result, the exporter's industrial and agricultural sectors can lose local and foreign customers. The loss may not hurt until the boom subsides and the flood of oil revenue turns into a trickle; the exporter's economy is left with industrial and agricultural sectors that have atrophied. This is known, in economics, as the Dutch disease, named after the decline of Dutch industry in the 1960s in the wake of an influx of revenue from the sale of North Sea natural gas. One remedy, economists have realized, is to "sterilize" oil revenues by keeping them offshore—investing a chunk of them in foreign stocks and bonds, for example. But a government that is mismanaged, greedy or just in desperate need of funds will let the money rush in. The Dutch economy recovered, but others have not been so fortunate.

Nigeria is like a specimen exposed to multiple diseases. Legions of young men, turning away from hard and low-paying farmwork, migrated to the cities for the easier jobs they thought would be available there. The jobs weren't there—the oil industry is not labor-intensive, and the Nigerian government, even if it hadn't lost funds to corruption and waste, did not have enough oil revenues to pay for the infrastructure projects that would put such a large labor force to work. Instead, the migrants coalesced into an urban underclass, Dickens gone to Africa. Some used their financial and language skills to perpetrate Internet scams—Nigeria is the origin of many of the too-good-to-be-true e-mail offers that fill in-boxes across the world. Ryszard Kapuściński, the Polish writer, noticed a similar abandonment of reason in Iran during the times of the shah. "Oil kindles extraordinary emotions and hopes, since oil is above all a great temptation," he wrote. "It is the temptation of ease, wealth, strength, fortune, power. It is a filthy, foul-smelling liquid that squirts obligingly into the air and falls back to earth as a rustling shower of money. . . . Oil creates an illusion of a completely changed life, life without work, life for free. Oil is a resource that anesthetizes thought, blurs vision, corrupts."

Ironically, oil's impact can be harshest on the communities where it is located. Instead of becoming rich and moving to mansions in fancy towns, as the fictional Clampett family did in the 1960s sitcom *The Beverly Hillbillies*, the people of the Niger Delta became poorer, watching as their land and water become polluted by an industry they did not own, had no control over and derived almost no income from. In the delta, once a vibrant marine habitat, fish died off and crops wilted. There was little compensation. Oil revenues that weren't stolen went directly into the national treasury, because the government in the capital controlled the revenues. The ethnic groups in the delta were not powerful enough to get their way in national politics. The modest funds earmarked for local development were, for the most part, stolen by officials and chiefs before reaching the people who were supposed to be the beneficiaries.

Rebellion, in such conditions, is inevitable. Early on, in 1966, Isaac Boro, an army officer born in the delta, cofounded the Niger Delta Volunteer Service and declared a breakaway republic. His revolt was crushed in twelve days by troops who rushed into the delta on boats supplied by Shell. Soon after, an accumulation of discord—partly over oil, but also over religion, culture and ethnicity—led to a massive and unsuccessful war of secession, the Biafran war, which killed as many as two million people. A new generation of activism emerged in the 1990s, led by the charismatic Ken Saro-Wiwa of the Ogoni tribe, which lived where oil was first found and whose people were its first victims. Saro-Wiwa formed a popular nonviolent campaign against Shell and the repressive military regime that was its partner at the time. In 1994, as martial law was about to be imposed on his restive home region, Saro-Wiwa predicted, "This is it, they are going to arrest us all and execute us. All for Shell." Soon after, on the orders of General Sani Abacha, the military dictator, Saro-Wiwa was arrested and later hanged after a show trial. Investigations after Abacha's death several years later revealed that he'd stolen $4 billion in state revenues.

Asari's rebellion was a violent continuation of this history. It was low-intensity warfare that killed thousands of combatants and civilians every year, and it had a postmodern touch, because helicopter gunships were pitted against militiamen who wore bullet-stopping amulets (or

so they believed). For the 30 million unfortunate souls in the delta—
the country's total population is nearly 150 million—life had become a
hellish vision that was part *Mad Max*, part *Waterworld*, and, with the
prevalence of adolescent fighters, a bit of *Lord of the Flies*.

Port Harcourt, in the fall of 2004, was its usual insane self. The power
grid was down, roads had holes the size of craters and foreign oilmen
were driven across town with armed bodyguards. Swindles and violence beckoned at every corner, and the police only made things worse.
With casually violent ways, paramilitary police teams were known as
"Kill and Go," because that's what they did. There was only one event
to be thankful for, and that was a truce between the government and
Asari's militia, the Niger Delta People's Volunteer Force. It was
rumored that Olusegun Obasanjo, the president at the time, had made
an offer that was too lucrative for Asari to turn down—a significant
sum of money for some sort of disarmament. In Nigeria, disputes
tended to be settled with guns or cash, and in this instance, cash had
done the job. Asari moved freely in Port Harcourt, residing in a small
hotel that was a thirty-minute bumper-to-bumper drive from mine (or,
as the crow flies, a mile or two).

When I arrived to talk with him, several dozen young men were
gathered in the hotel driveway and bar, wearing the uniform of toughs
at rest—loose T-shirts and sweatpants. They drank Star beer, even
though lunch was hours away. At this moment, they were interested in
nothing more challenging than watching soccer on the bar's television,
but they were not incurious, at least if their nicknames were windows
into their minds. Nigerians love nicknames, and the young men I
encountered at the hotel told me to call them Justice or History or
some such moniker. It is strange to meet a teenager who introduces
himself by saying, in literal truth, "I am Handsome!"

I was led to a small suite where a klatch of these youths were watching a DVD of Asari speaking to villagers in the Niger Delta. They
watched avidly, because Asari was a spellbinding orator, and they failed
to notice that their leader had woken and shuffled into the back of the

suite from the adjoining bedroom. Asari was wearing gym shorts, a T-shirt and flip-flops. He watched the TV with the look of a groggy sculptor sipping his first cup of morning coffee and assessing the previous day's work. He seemed pleased.

On the screen, a village elder lamented the abundance of oil in the ground and the lack of food in people's stomachs. Behind him, a boy peeked through the crowd, wearing an Oakland A's cap.

"Brothers and sisters, your salvation is in your hands," Asari, on TV, told the villagers. "Today you must choose whether to be free or in jail. If you want to be free, say yes."

The crowd roared its assent.

A cell phone rang in the suite. It was Asari's, and he told the caller to phone back later. He returned to watching the best entertainment going.

"Now is the time for us to fight together," Asari told the crowd. "On this road there is weeping and the cracking of teeth, but the only way we can win is by fighting. We will fight until the enemy abandons control of our resources."

There was more cheering, and at the speech's conclusion Asari was driven away in a white Hummer, standing up and waving through its sunroof like the grand marshal at a New Year's Day parade.

It was easy to understand his allure. Asari possessed the cadences of Dr. Martin Luther King Jr. and the dramatics of Fidel Castro, and he borrowed lines from Bob Marley. His words inspired not only the crowd in the delta and not only the young men watching the DVD— their keen interest was not fabricated—but even Asari himself. Sleepy when he'd padded into the suite, he was now energized. As one of his followers cued up another clip, Asari tapped my shoulder and led me down a corridor to a room that was under renovation or had been relieved of its contents by his followers.

His eyes jumped to the digital audio recorder I placed at his side.

"Can I buy it from you?" he asked.

It would not be wise to refuse a warlord.

"You can buy it on the Internet," I suggested.

"No. I need it now."

I said that I needed it to conduct my work in Nigeria. He understood. The interview began.

Asari was an integral part of Nigeria's ordered chaos. That is an
odd phrase, "ordered chaos," but it gives appropriate due to the connections among the men and women for whom the country is a giant
stage where they fight and bargain for the treasure under their feet. In
fetid swamps, crowded slums and corporate towers, there is a form of
nonstop interaction between warlords, governors, tribal kings and oil
executives, like actors and stagehands in the same production. The
ones I met would move on sooner or later—to another position,
another place of exile or a grave—but others would replace them,
because this was not a theatrical production but a system of perpetual
conflict that compromised everyone involved in it.

Asari was one of six children in a middle-class family; his father was
a judge, his mother a housewife. He attended a Baptist secondary
school, became a Marxist in college, dropped out of law school twice
and failed on two runs for elective office. He had converted to Islam
before his electoral career, changing his name (good-bye Dokubo
Melford Goodhead Jr.; hello Alhaji Mujahid Dokubo Asari), and
became a Ramadan-observing militant. He professed admiration for
Osama bin Laden as well as Nelson Mandela. Because personal as well
as moral flexibility is required in tumultuous Nigeria, where life proceeds with the complexity of chess rather than the reliability of check-
ers, Asari was not accused of inconsistency.

Nigeria's population is the largest in Africa and is composed of several hundred ethnic groups divided into an even greater number of
tribes and subtribes with their own dialects and disputes. Asari
belonged to the delta's largest tribe, the Ijaw, whose members are pre-
dominantly Christian. He was one of the tribe's few Muslims, but in
the rock-paper-scissors game of identity in Nigeria, tribe trumps religion; the Ijaw do not care if you worship Allah or Apollo—you are
theirs. In the late 1990s Asari was involved in the creation of the Ijaw
Youth Council, and became president of it in 2001. A slogan was
adopted, "Resource control and self-determination by every means

necessary," and that meant war. Asari went into the creeks and formed his militia.

It was a slow day at his hotel-turned-headquarters, so we conversed with few interruptions, though once, in response to an argument in a nearby room, Asari shouted something in Ijaw and the dispute quieted. Asari did not look like a warlord. He was of average height and chubby, with the eternally fatigued bearing of a man who must work several jobs to stay afloat. Yet he was the real thing. When I noticed a scar on an arm, he displayed others that were the result of combat, though he could not always recall the injury. "This one," he said, touching a three-inch tear in the flesh of his leg and pausing over it, "I don't remember how I got it."

Asari was fighting an alliance of the foreign companies that extracted the delta's oil (Shell drilled the most) and the central government, which hired them and provided security for their operations. It was the same alliance that Saro-Wiwa and Isaac Boro had fought against. For them, the government and the companies were partners in a regionwide, decades-long crime. Of course, Asari's fighters were also receiving money from the oil companies, in the form of ransoms paid for workers taken hostage or protection money to help defend oil installations from . . . their own attacks. It was a closed loop of recirculated violence.

"The oil companies are working with an occupation government that does not have a legal right to the resources," Asari said. "They should pack and go because they have contributed so much to the deprivation, oppression and suffering of our people. There is oil in Alaska, there is oil in Siberia. They can explore for oil there."

Asari liked to be known as a freedom fighter—that's why he showed off his scars—and he offered a clear definition that explained his collision with the government in Abuja.

"Freedom means the restoration of sovereignty to our people," he said.

Sovereign, as in apart from Nigeria?

"Yes, as a separate nation. We are not going to be apologetic about it. We have the right to self-determination."

He responded without anger when it was suggested that he might fare no better than Saro-Wiwa, Boro or, for that matter, any rebel who wishes to separate his region from a government that is brutal, well armed and determined not to let go of a territory under which a treasure is buried.

"Like the people of Ireland and the people of Chechnya and many other people, if it is not the wish of God that I should be the one who succeeds, a better person will take over," he said.

It was impossible to know whether Asari would reach the promised land that had eluded his predecessors, but one thing was sure: Nigeria was collapsing. Countries do not rot as fast as men or buildings, but rot they do. Laws devolve from tools of order into excuses for payoffs. You need a permit? Very good. How much will you pay? In parts of Nigeria, state control was an idea rather than fact. This is a hallmark of failed states: they fail to even exist. In the delta, there was no formal or informal demarcation between areas controlled by the government and those held by the rebels. Asari, at the time of my visit, was the delta's alpha rebel, so I needed travel permission from him, not the police, the army or Shell.

He assigned a young aide, whose nickname was Senator, to take me upriver. The next day, we traveled into the delta and wound up on a leaky canoe in a dying river with a tearful king.

The global distribution of oil is wildly irregular. Lots in Russia, not nearly as much in China. Vast reserves in Venezuela, none in Guyana. Large amounts in Libya, minor quantities in Egypt. The unevenness applies within countries, too. America's remaining fields are in a handful of states, yet even in those, there are geographic gaps. The oil of Texas is lopsidedly concentrated in the west and east of the state. Though we think of Saudi Arabia as swimming in oil, in fact, most of the country has little; its reserves are concentrated in Ghawar, the 174-mile-long field discussed in chapter one. The same for Iraq—it has two giant concentrations of oil, one near Kirkuk, the other near Basra. Elsewhere in the country, relatively little. Geology is not destiny—Texans are not at one another's throats—but the uneven distribution of oil can exacerbate regional tensions. The Shiites concentrated in Saudi Arabia's eastern oil region would prefer autonomy, if not independence, but the Sunni royal family has used its security forces to prevent those sentiments from going anywhere; also, the region where oil was found has been repopulated with Sunnis. In Iraq the huge oil reserves located around Kirkuk could be the setting for a war between Kurds and Arabs, both of whom claim ownership rights over the area and its oil.

Today, you needn't be a Marxist to be interested in the role of natural resources in political conflicts. One of the most prominent experts in this hybrid field is Paul Collier, an economist formerly at the World Bank who coauthored a study of 160 countries and 73 civil wars since 1960. The report was titled "Greed and Grievance in Civil War," and this is how Collier described its findings: "Dependence on primary commodities substantially increases the risk of conflict, unless the primary commodity is extremely plentiful, such as in the case of oil in Saudi Arabia. In a country with no primary commodity exports at all, the risk is about one percent in a five-year period. In a country with high dependence on primary commodities, which means about 30 percent of its national income comes from primary commodities, the risk is around 23 percent." A key problem is political maturity. If a country has a stable and open government and a sense of national identity, like Norway, it has a high probability of resolving the disagreements over who owns the oil and how revenues should be distributed. But Nigeria, like most countries, lacked those attributes.

The Niger Delta was a territory of 100 percent risk. With Senator guiding the way out of Port Harcourt, we drove for nearly two hours to Abonnema, where the road ended, and got into a canoe with an outboard engine that conked out several times as we made our way deeper into the delta. The delta is one of the world's largest wetlands, its ecosystem once nourishing almost any living thing that might have walked, flown, swum or wiggled onto Noah's Ark. If the delta had been blessed with no oil, it might now be a wildlife sanctuary in which Western tourists would pay hundreds of dollars a night to fall asleep in ecolodges listening to the sounds of herons in the trees and manatees in the creeks. Fate has not been so kind. Our first stop, after forty-five minutes of navigating a sweltering maze of muddy, mangrove-lined creeks, was Tombia, a town attacked on multiple occasions and by multiple forces, in 2004.

Tombia was a shambles, half its homes burned or bombed beyond repair. As the canoe glided onto the riverbank, a dozen survivors came out, and their manner was not warm. They were young men, fighters, some with soiled bandages. Fingers and hands were missing; limbs were swathed in gauze that was caked in pus. I stayed by the canoe as they argued with Senator. After a few minutes, Senator came to my side and whispered, "We'll probably have to go."

The puzzle was not hard to piece together, because I had heard of Tombia's sufferings. In the delta war, the town had sided with Asari, who sheltered there from time to time. A rival militia led by a warlord named Ateke Tom had attacked and seized Tombia. Ateke Tom had taken a shortcut to the position of rebel leader: much of his funding and weaponry allegedly came from the local governor. This is often the way things go in collapsing states. To undermine a semipopular rebel like Asari, a government, realizing that its own forces are inept or too unpopular to win a hearts-and-minds struggle, funds a new rebel force that it quietly controls. If the new one becomes too popular and uncontrollable, yet another one is created to take on the last. Antigovernment agitation becomes a perpetually splitting atom.

After retreating from Tombia, Asari's fighters regrouped and reconquered it. The government then stepped in. Army forces attacked in the brutal way they usually do, with helicopter gunships strafing anything that moved, while speedboats landed soldiers, who shot and looted their way through town. A dozen people were reported killed, and the town's population was too frightened to return—but in any event, there was not much to return to.

The young men who'd argued with Senator were upset that Asari had failed to compensate Tombia for its misery; the boss was living well in the city while his followers were perishing in the delta from their suppurating wounds. Senator—who, fortunately, had the charm of a senator—eventually soothed these survivors, and they provided a tour of the remains of their town, which had had a population of several thousand before the attack. The youths' leader, whose nickname was Prince, led the way, pointing out the destruction with the stump of what used to be his right hand. Even the Lutheran cathedral, St. Stephen's, had been destroyed, a fire consuming its walls and roof. Its timid pastor, living in a shack and shivering from malaria or fear of the bitter youths who ruled this wasteland, said it had been constructed by British missionaries in 1915. A sign by the church declared, "Tombia is dedicated to God. Jesus the King over the land. Holy ghost in charge."

A boy who looked to be twelve years old and was blind in one eye stood in front of a house that had burned to its concrete foundation. His older brother had been killed, he said, his town was now dead and his river was dead too, due to oil pollution. He could not possibly catch enough fish to nourish himself and his dead brother's family. He was angry and hopeless at the same time; the result was listlessness. The government, the army, Shell, Asari, Ateke Tom, the writer who would leave in a few minutes—they would not help. His only hope was, it seemed, the holy ghost.

Back on the river, the journey continued into a landscape of ruin. The local man steering the canoe did not consult a map as he navigated a maze of creeks; he knew the way to wherever we were going. At times, we ducked to avoid low-hanging vines and branches. There was a primeval feel to this passage, though every few minutes a sign along the shore said, "Warning: Gas Pipeline Under Pressure" or "High Pressure Gas Pipeline: Do Not Anchor." Intertwined pipes emerged from the water in a tangle of metal that had the appearance of industrial art; these were wellheads, from which the flow of oil was controlled. At some spots, the shoreline was shaved of vegetation and fenced off, to protect flares and pits that burned off excess oil and gas. The earth in these places was, quite literally, on fire.

Most wells in the delta were run by Shell, which hired subcontractors for some work. That's why a drilling barge came into view that bore, on its flank, a familiar name—Halliburton, the energy-services company that was once the home of former Vice President Dick

Cheney. The barge was painted in Halliburton's colors, red and white, and was guarded by several dozen soldiers, who stared coldly as Asari's people passed by. It was understood that war would return soon.

It was time to visit a kingdom that even smelled of rot.

The canoe headed to Oru Sangama. The village's defining feature, apprehended on first inhalation, was a heavy odor of sewage that had fused with humidity to form a fecal mist. Sangama's several hundred residents relieved themselves in a creek that was just a few steps from their homes; the creek was dead, or nearly so, as was the sickly mangrove jungle around it. The villagers had no clean water, no electricity, no school, no doctor or nurse, not even any wildlife in the trees. The walls of their homes were made of mud bricks, the roofs of sheet metal. The good news was that these homes—several dozen in all—were new, but only because the army had destroyed the old ones in a recent assault.

I was led to an open-air hut and offered a plastic chair and a soft drink. Someone said "His Highness" was busy at the moment but would arrive shortly. To make me feel at home, a small generator was switched on, along with a television connected to a satellite dish. A villager tuned in Fox News. Thanks to the TV, the residents of Sangama were familiar with Bill O'Reilly and the enviable standard of living enjoyed by the Americans who consumed the oil that was killing them off. About 40 percent of Nigeria's oil exports went to the United States. From the wellheads I saw in the creeks, the oil flowed to processing stations and then to export terminals like Bonny Island, where it was poured into storage tanks before being loaded onto supertankers and shipped to American ports, where it was offloaded at coastal refineries and turned into the gas that kept our SUVs moving.

When the TV became quiet, a steady roar, like a giant flame-thrower, could be heard. About a thousand yards away, across the fetid creek, stood the Soku natural gas plant, which was part of a multibillion-dollar network of facilities collecting natural gas from Shell's oil fields. The village was in the shadow of Soku's main tower, which was several hundred feet tall and shot into the air a plume of fire;

the natural gas, burning furiously, sounded like a rocket launch. As darkness fell, Sangama became illuminated by the flare's reddish glow; the town remained lit in this fashion until the sun rose in the morning. The Martian tint was nearly bright enough to read by, but it was entirely unwanted, because it was lethal.

This flare was one of hundreds I'd noticed while flying into Port Harcourt; they were so large and bright that you could see them from an altitude of thirty-five thousand feet. They existed because when oil is brought to the surface, natural gas rises, too; in the industry, it is referred to as "associated gas." Oil firms have several options for dealing with it: transport the gas to customers elsewhere, reinject it into the earth or burn it off. The first option is feasible if the volume of gas is large enough to justify pipelines or liquefication facilities; these often cost billions of dollars. Reinjection is environmentally safe, but the technology, though less costly than pipelines, is still expensive.

Flaring is the cheapest way to deal with the problem, though its environmental and health costs are colossal. The carcinogens that flaring releases include benzene, benzopyrene and toluene. The metals emitted include mercury, arsenic and chromium (the contaminant that

Gas flares in the Niger Delta

was publicized by Erin Brockovich in California). The released green-house gases, which cause global warming, include carbon dioxide and methane. Emissions of sulfur dioxide and nitrogen oxide are so severe in the delta that acid rain eats through sheet-metal roofs; they need to be replaced every few years, though the Nigerian army hastened the pace at Sangama.

Illnesses caused by flaring include renal and cardiovascular failure, cancer, leukemia, emphysema, bronchitis, immune-system dysfunction and reproductive disorders. Yet it is difficult in the delta to know the toll. Epidemiological studies have not been conducted because funding is not available and the delta has become too dangerous for researchers. When someone in the delta dies of thyroid cancer or leukemia caused by gas in the air or oil in the water, they are buried without an autopsy. These deaths, and certainly the causes of these deaths, are the prover-bial trees that fall without being heard.

This does not happen in developed countries, where oil firms are obligated to invest in technology and infrastructure that reduce flaring to almost nothing. The United States, for instance, burns less than 1 percent of its associated gas. Flaring was banned in Nigeria in 1984, but the practice continued because the only way to stop it, when pipelines or reinjection were deemed too costly, was to cease bringing oil to the surface. The Nigerian government, addicted to the profits of oil, did not order that. As a result, nearly 55 percent of the country's associated gas was being flared when I was in the country. According to a joint report by the World Bank and UN Development Programme, Nigeria was burning 2.5 billion cubic feet of gas every day, releasing about 70 million tons of carbon dioxide per year; nearly 20 percent of the gas flared in the world comes from Nigeria. Shell claims to have reduced flaring, and the Soku plant was, actually, part of its effort—the facility processes and transports associated gas for use by consumers and also reinjects gas into the ground. But the gas coming out of the ground was more than even Soku could handle. The deadly excess burned.

Across the creek, Sangama consisted of a few dozen huts crammed into an area the size of a football field. Privacy did not exist; if a husband and wife quarreled, the village heard it. Watching TV in the red glare cast by Soku's flare, I heard, in the surrounding huts, young children cough the evening away. They were sick, and if associated gas was the culprit, the crime continued without punishment.

"His Highness is coming," someone whispered into my ear.

King Tom Mercy strode into the hut, wearing a T-shirt, pants and a frown. He was the forty-two-year-old leader of an Ijaw subclan whose dominion extended over a dozen villages and a few thousand souls. This was size enough to warrant, in the delta, the most monarchial of titles. For informal interactions, villagers used their leaders' royal titles as surrogate first names, as in "I think Highness is napping," or "King, please pass the salt." The king's brother, who went by the nickname of Daddy, was introduced as Commander in Chief.

The brothers let another man, the community development secretary, tell the village's story. He recited it in Homeric fashion, half-singing, gesturing broadly, his voice rising and falling operatically. Sangama was founded long ago, he said, but was abandoned in the early 1990s due to intertribal clashes. Resettled in 1999, the village got little compensation during the construction and expansion of the Soku plant, literally a stone's throw away. A large generator had recently been delivered by Shell, and a concrete shelter built for it, but the generator would consume more fuel than the village could afford, so it remained wrapped in plastic by the creek, rusting by the hour. As a gift, it was akin to a donor giving crutches to a man who is paralyzed.

If Shell truly wished to provide electricity, it could have laid a cable across the creek to connect Sangama with Soku's power system. At Soku, where several hundred workers lived behind a wire fence patrolled by soldiers and private security guards, there was an abundance of electricity, as well as running water, a medical station, a helipad, showers, telephones, air-conditioning, hot meals and so on. The facility had its own power plant. It was a high-tech fortress of a sort that existed across the country. One of the best photos I have seen of Nigeria is an overhead picture of a Shell oil terminal at Bonny: the facility is pristine and well lit, with green lawns and smooth roads, yet

Bonny oil facility in the Niger Delta

on the other side of a river is a decrepit slum with as much misery as there are barrels of oil on the opposite bank. The poignance is greater than any picture can show, because the Niger Delta is stained with dichotomies of this sort, and one of them is fittingly located at Chevron's terminal at Escravos. Beginning in the 1400s, Escravos was a departure point for the West African slave trade. In Portuguese, *"escravos"* means "slaves." Nigeria's oil, waiting to be shipped to America, was housed in far better circumstances than Nigeria's people, who were going nowhere.

His Highness was not pleased.

"That gas plant is the largest in Africa," King Tom said, cutting off his subordinate. "This is where the oil and gas comes out. They could give us water, give us light, give us scholarships, give us jobs. We would not quarrel with anyone again. We have tried everything, used lawyers and dialogue, and we see there is no way. The next thing is violence. We don't care if everyone dies, we will burn it."

The next morning, he offered a tour of his domain. He was splendidly attired for the occasion, with a silver-tipped cane, gold bracelet, emerald brooch and black hat. His disposition was the opposite of splendid, however. As his twin-engine canoe slid from Sangama, he remarked, "They are treating us like goats, not even human beings."

The canoe moved through mangrove creeks in which there was no screeching of monkeys, no hippos or crocodiles in the water, no butterflies floating in the air. I began counting the number of birds, because wetlands are usually filled with them. I noticed one, a white egret, but not another until five or ten minutes later. I kept track of the time between bird sightings; they were never more frequent than one every few minutes. In these wetlands, the wildlife was all but gone. Between the war and the pollution, it seemed to be both a killing zone and a dead zone.

We found a fisherman using a pole to propel his leaky dugout. He had been fishing since the previous day but had not caught nearly enough to feed his family. In his canoe, there were a half dozen fish the size of large minnows. "Wherever I go, there is pollution," he said. "All the fish have gone to the ocean." His gear consisted of a hook at the end of a string that was attached to a stick. He could not afford a fishing rod.

This journey required, for comprehension, the imagination of a science fiction devotee. We passed a small island that was known as Little Russia. The origin of its name was not clear, but the island served a distinct purpose—it was where prostitutes lived, servicing the needs of soldiers and oil workers. On its shore, young women stood in the shade of shacks fronted with empty beer bottles and off-kilter picnic tables. A sign over one of the shacks announced, "Warri Plaza—Food Is Ready." The girls waved.

We saw pits of burning oil and we saw flames roaring from flares on the ground; the earth was hissing fire. The smell of oil was strong in these places, and even when wells or flares were not visible, the odor of petroleum was present. Where did it come from? I looked down and saw a film of oil on the river. According to official statistics, between 1976 and 2001, there were, on average, more than five spills a week in Nigeria, but according to unofficial estimates, the true figure could be ten times higher. Shell and other firms claim to abide by first-world

King Tom Mercy

standards, but that seems a fairy tale, the punch line to which was announced by a sign at a flow station that was dripping fluids into the water: "Keep Nigeria Safe and Clean."

The canoe stopped in front of six leaky wellheads coated in oil that fell, drop by drop, into the water. King Tom said Shell planned to build eighteen more wells and two more flow stations in this area.

"How can we expect to catch fish?" he asked.

His anger was no performance.

"Let's go," he ordered.

We soon passed a patrol boat with unsmiling soldiers.

"You see how we live."

Oil brought more than economic and environmental misery to Sangama.

A month before my visit, the army had attacked the village in much the same way it had attacked Tombia, and with a similar result. Helicopters and speedboats all but destroyed it, though soldiers, before torching the better huts, made of wood, looted them of anything that merited looting, including pots and pans. Most villagers got away, so just two were killed, the king said. The assault was unprovoked, he added, and he accused Shell of complicity. The company had evacuated personnel from Soku hours before the attack—the villagers had seen helicopters ferrying employees away. The company knew something was going to happen.

The king was not as unaware as he pretended to be. Like Tombia, Sangama had been used as a base by Asari; King Tom was a supporter. On the surface, the situation seemed odd. Why would Asari and King Tom, who opposed Shell and all it stood for, live in the shadow of a gas flare rather than attack it? The answer revealed a perverse economic dynamic that ensured violence but discouraged total war. All sides shared a mutual interest in avoiding a long-term closure of the industry because all derived income from the flow of oil—not just Shell but also the rebels who tapped the pipelines and bribed the military to look the other way. If Asari or King Tom wanted to shut down Soku, a rocket-propelled grenade fired across the creek would do the job. Sangama was a dagger to the throat of Soku that nobody wanted to use. This was a Nigerian rather than a Mexican standoff.

Though a few burned buildings remained as reminders of the army attack, Sangama was being rebuilt quickly, and this was a hint of its participation in "bunkering." That's the word Nigerians use to describe the siphoning of oil from company pipelines. Throughout the delta, bunkering deprived Shell and other companies of about 15 percent of their output, which meant billions of dollars a year for the bunkerers. The theft was reasonable to the locals, because if King Tom and Asari's other allies didn't seize oil for themselves, the government and Shell would take it all and return little in the way of services or aid. Yet bunkering seemed to pay primarily for guns and bullets for the fighters, and the occasional rebuilding of villages the military attacked, rather than food and medicine for the people. Bunkering was a symptom of the problem, not an answer to it, because despite its bunkering income, Sangama was a miserable place. Just as corruption made government oil revenues disappear, it appeared to do the same with a sizable portion of the rebellion's take.

Even the military got a cut. Asari had to ship the purloined oil out of the delta, and this could not be done so easily, because oil barges are not like gems that can be slipped into a pocket for smuggling purposes. It was an open secret that military and police officials were paid to look the other way as illicit barges moved along the delta's waterways. Two navy admirals were even court-martialed for their involvement in bunkering in 2005. (The surprise was that they were court-martialed, not that they were involved in bunkering.) It *is* confusing—were the military and rebels fighting against each other or doing business together? The answer was . . . both. Army assaults might be mounted for traditional military purposes, or to punish one of Asari's subordinates for failing to pay sufficient tributes. This may have explained the attack on Sangama—perhaps King Tom had been slow to share his bunkering revenues with local military officials. King Tom wouldn't say; he stuck to the attacked-out-of-the-blue script.

At the end of the day, it was hard to make sense of what I saw. I slept in a mud-walled room in Sangama, sharing a bare mattress with Tony Iyare, a Nigerian journalist traveling with me. We shared the mattress because its grime was less offensive than the floor's. Between the humidity, mosquitoes, fecal stench, red glow, gas fumes and coughing children and the novelty of sharing a filthy bed with a traveling companion who was six foot four, it was a night of unsound sleep. I understood why Senator had said, earlier in the evening, that Nigerians cannot think straight anymore.

At a bend in a river, Shell was trying to make amends.

The company's involvement in Nigeria was older than the country itself. In 1938, when Nigeria was still a British colony, a Shell subsidiary, Shell D'Arcy Exploration Company, was granted exclusive exploration rights. Commercially viable deposits were not found until 1956, after which Shell reached a fifty-fifty profit-sharing contract with the colonial government. A few years later, Nigeria's newly independent government awarded exploration contracts to other firms, including Mobil and Texaco, but Shell retained control of the largest number of wells. As the government took a more active role in the 1970s, foreign companies were required to transform their wholly owned operations into joint ventures with state-owned firms. In most of those ventures, foreigners were the "operators," meaning they managed the enterprise. That's one of the reasons foreign companies, rather than their government-controlled partners, became such lightning rods—they were the ones who were on the front line, so to speak.

If the Nigerian government had been efficient and fair with the revenues from its share in the joint ventures (in addition to the taxes it received from the ventures' sales), Shell might not have faced so much wrath from the public. Initially, in the 1960s, half of the government's income from the joint ventures went to the oil-producing states in the delta, with the rest going to the federal treasury. But in the 1970s, after the failed secession attempt of the Biafran war, the laws were changed and everything went to the federal government. The funds that were then handed back to the oil states were not particularly generous and were often stolen by local officials before they could reach their constituents. The formula changed a decade ago, with more money going directly to the oil states, but the sense of grievance has not changed—the people in the delta feel cheated.

In the 1990s, multinationals became aware of the downsides of being unloved not only in the third world, where violence was becoming a default response to their misdeeds, but also in Western markets, where consumer boycotts punished their bad behavior. Shell faced a public relations disaster after the execution of Ken Saro-Wiwa because the company was regarded as having encouraged the military government to eliminate the human thorn in its balance sheet. Saro-Wiwa's murder led to anti-Shell boycotts in America and the United Kingdom, and in response, Shell devoted a larger (but still modest) share of revenues to development projects. The king wanted me to see one of the latest ones.

The king's canoe neared Elem Sangama—a sister village to the king's Oru Sangama—and the first impression was of a paradise in the delta. Elem Sangama had a water tower, school buildings, a health clinic, several dozen market stalls, a town hall, electricity poles and paved walkways. It was a Hollywood-ready example of what a thriving African village was supposed to look like, at least from afar, and this was thanks to Shell, which paid for all of the construction. It was the sort of project oil firms publicize in annual reports and advertisements about their roles as good corporate citizens. If only there were more of these projects, one thinks, there wouldn't be so much violence in the delta.

The reconstruction of Elem Sangama was yet another lesson in rot, because almost everything Shell built was unused and unusable. The water tower and the pipes connected to it were dry because pumps had never been installed. The market stalls were empty because the village did not have enough goods or money to justify a market; the only commerce was conducted in a tin shack selling beans and beer, and the shack, a few steps closer to the huts where the villagers lived, had not been built by Shell. The health clinic was empty and padlocked because Shell had not provided medical equipment, medicines or, for that matter, a doctor or nurse. The electrical lines were filled with electricity only occasionally because Shell had donated a gas-powered generator but no gasoline (the same problem as in the king's village). A village leader who showed me around was unimpressed when reminded that Shell executives said it was not their responsibility to provide everything the village needed.

Street in Port Harcourt, Nigeria

"But is it their responsibility to take our resources?" he said.

Shell's upgrades included signs for the footpaths it had paved, although a poor village in the Niger Delta was not in need of asphalt paths or painted signs for them. It was an almost mocking touch that the path in front of the health clinic was called SPDC Road, which stood for Shell Petroleum Development Company, and it was stranger still that a sign on the waterfront said "Sea Side Lane," evoking a vista of blue water, whereas the actual waterway was an oil-ridden creek. The signs were announcements of Shell's dominion and disregard.

Before leaving Nigeria, I made an appointment in Lagos, the former capital, with Chris Finlayson, who, depending on one's point of view, was a liar, truth teller, environmentalist, polluter, puppetmaster, puppet, devil in pinstripes, honest executive—the list continues. Finlayson was the director of Shell's operations in Nigeria.

Shell's headquarters, in the center of the city's business district, had a moldy look, an organic consequence of an equatorial climate that accelerates decay in cement and flesh. Finlayson's upper-floor office had excellent views that highlighted the distant horizon but not the dreadfulness of life below. Finlayson, whose physique is well-rounded, had the congeniality of a decent man who'd won a modest lottery. He motioned to a conference table; there we joined his spokesman, whose cell phone interrupted us with a rap song about murder. The get-acquainted chatter touched on a photo on one wall that showed a small passenger jet. Finlayson moaned theatrically, saying it was one of many Shell used to fly its employees around the country. The company maintained its own fleet because commercial airlines in Nigeria were unreliable and unsafe. The cost of operating these planes, Finlayson groaned again, was enormous.

Shell presents itself as a saddened bystander to social collapse. The company and its executives often stress their regret over the misery and their desire to make things better. To an extent, their concern is genuine, as is their frustration, because their development efforts can *incite* violence, as communities that do not receive aid become jealous of communities that do. While in Lagos I was told of a health clinic in the delta that Chevron had built and equipped. It was burned down by an adjacent community. Chevron rebuilt the clinic, and it was burned down again. There was no third effort.

A cascade of reports over the years has shown that good-faith development efforts are overshadowed by day-to-day practices that have helped make the delta as violent as Chechnya and Colombia. One such report, entitled "Peace and Security in the Niger Delta," noted that Shell's facilities and operations relied on the protection of law enforcement agencies that used "jungle justice," which means murder and torture. Shell was also criticized for engaging in bribery by awarding no-work contracts to front companies owned by local leaders. "The manner in which [Shell] operates and its staff behaves creates, feeds into or exacerbates conflict," the report states. "After over 50 years in Nigeria, it is therefore reasonable to say that [Shell] has become an integral part of the Niger Delta conflict system."

The report was remarkable not for what it said, which was a standard critique of the company, but for who said it: Shell. The company, wishing to understand what had gone wrong and how to correct its problems, had commissioned a confidential study, which had leaked out shortly before I arrived in the country. The ninety-three-page document confirmed everything the company had denied or avoided for years—and then some. In the wake of the report's publication, the task of damage control fell to Chris Finlayson.

"Ninety-eight percent of what was in that report was good stuff," Finlayson said. "The one negative point"—he was referring to a prediction that violence may force Shell to abandon its wells in the delta—"we probably don't agree with. But we accept that we have to improve. I'm not going to dissemble. We accept that we can always do better, and that's what we are trying to do."

This is what is supposed to happen—a company acknowledging and correcting its errors. But Finlayson's we're-sorry-and-we've-learned-our-lesson narrative has been repeated by Shell executives since the killing of Saro-Wiwa in 1995. My visit to Sangama had introduced me to a fresh example of the company's dissembling.

I reminded Finlayson of the army attack, just a month earlier, on

King Tom's village, adjacent to the Soku gas plant operated by Shell. King Tom had told me that Shell provided money, food and other support to soldiers who guarded Soku and who were involved in the assault on his village. I mentioned to Finlayson that this would seem to align Shell with a brutal military attack.

Finlayson responded quickly.

"We do not pay for troops, we do not provide any arms or other lethal support. It is not right for commercial companies to do so."

Does that mean a ban on "nonlethal" support for troops?

"We do provide accommodations and we do feed troops when they are protecting our operations. We do obviously request protection where we feel our operations are under threat."

It seemed an odd distinction, to pay for everything but the soldiers' bullets and then deny any responsibility for violence committed by the soldiers. Finlayson knew this was an awkward position for a multinational that wished to be viewed as politically neutral. Like any multinational, Shell would like to be known as a good corporate citizen, not as the funder of military pogroms.

Finlayson's hale demeanor had gone somber. I mentioned that King Tom had said that company helicopters had flown into the plant to evacuate workers a few hours before the attack. If Shell had advance knowledge of the attack and had failed to warn civilians who were in the line of fire, its silence would seem to prove, once more, the king's charge that Shell was an accomplice to the army attack.

Finlayson selected his words slowly, as though each one was run through a filter before leaving his lips.

"We had intelligence that government activity was increasing in the area. We had no idea where the activity was going to be, but we knew that the area around the gas plant was at risk. We took the action of protecting our own staff and flying people out. But we don't know what the military are going to do, we don't know where they're going to do it."

The Soku plant, which was part of a multibillion-dollar gas project, consisted of high-pressure pipes and tanks filled with flammable and explosive materials—mainly natural gas. It is difficult to imagine that the army would not let Shell know of an imminent attack that would present grave risks to the Soku facility and the workers there. And it is almost unimaginable that Shell would not contact the army to ask about rumors it had heard of military activity that would involve the detonation of a considerable amount of munitions along the fence line of one of the largest gas installations in Africa.

I rephrased the question.

"The military did not tell you ahead of this attack to do what you needed to do to evacuate the plant?"

The voice of Shell replied firmly.

"No, definitely not."

Once it has started, rot is hard to stop, whether in a body or a nation.

The truce in late 2004 allowed Asari to live openly in Port Harcourt, where he even had a police escort when he went about town. Asari told me it would not last, and that Nigeria would get more violent, sicker. He evoked South Africa when it was ruled by a minority white regime. "I've been to South Africa," he said. "I have been to Soweto. Everywhere I went, I said to South Africans, 'We can switch positions. Apartheid was one hundred percent better than what is going on in Nigeria.'" Asari was known to wield a fine quote, but his point was germane. Nelson Mandela was freed by his white jailers. Saro-Wiwa, the delta's emancipation leader, was hanged by his Nigerian captors. And less than a year after Asari warned me that worse was to come, security forces arrested him on charges of treason. His followers responded by occupying oil facilities and warning, in a particularly vibrant statement, of "grave mayhem" in which their enemies would be fed to vultures. A new rebel group announced itself, the Movement for the Emancipation of the Niger Delta (MEND), and embarked on a campaign that was far more violent than the one led by Asari. In MEND's first months, more than sixty Western oil workers were kidnapped. Car bombs, which were all but unknown in Nigeria, were detonated against military and oil targets. To placate the rebels, the government eventually released Asari from jail. The violence continued, fiercer than before. Nigeria's decomposition seemed unstoppable.

4

Contamination

Until the drillers arrived, the Oriente region of Ecuador was an undisturbed rain forest inhabited by indigenous Indian tribes, mainly the Cofán, Huaorani, Secoya, Siona and Quichua. Tens of thousands of Indians were sprinkled through a lush expanse of trees and streams in an area as large as Rhode Island. No roads penetrated very deeply into the area, so the Indians lived in relative seclusion, except for occasional explorers, who did not all have the benefit of surviving their explorations. But in the early 1960s, the absence of roads did not stymie a new sort of visitor. These were American geologists who used helicopters to drop into the Oriente, and when they found what they were looking for the government awarded Texaco a twenty-eight-year concession to extract the region's oil.

The world offers a multitude of environmental disasters created by extractive industries that dig for oil, gold, silver or other minerals. Calling these events "tragedies" may not be right, because the word implies a course of events that went in an unexpected direction, like an early death, a sudden landslide, a plane crash. Mineral ecocides have happened often enough and predictably enough to be cast as the order of things. In countries too weak to control powerful industries that tend to behave responsibly only if they are required to, the invasion of bulldozers and other machines of extraction is a disaster foretold. The unexpected twist in the story of the Oriente is that an unprecedented lawsuit might provide a measure of justice.

. . .

Quito, the capital of Ecuador, sits atop the Andes, more than nine thousand feet above sea level. Driving east from the city, one passes a succession of high-altitude valleys that have the stark grandeur of an Ansel Adams photograph, albeit with an occasional llama. The winding two-lane road, with flimsy guardrails that wouldn't halt a tricycle, descends to the Amazon basin in ill-mannered serpentines. As the terrain flattens, the cloud forests of the Andes morph into a steamy infection of shacks, cattle, farms and people. Curving alongside the highway is a thick pipeline filled with petroleum that is used as an elevated walkway by kids and adults who don't want to get stuck in the mud along the road. They walk, literally and magically, on a path of oil.

This is the Trans-Ecuadorian Oil Pipeline System, known by its Spanish acronym, SOTE. More than three hundred miles long, it was built in the 1970s. In 2003 it gained a twin that doubled Ecuador's export capacity, transporting Oriente oil over the Andes to the Pacific port of Esmeraldas. Ecuador now produces 500,000 barrels a day, with the largest portion of exports going to California. Whether it is irony,

Oil spill in the Oriente region of Ecuador

parody or farce, one of the most environmentally conscious states in America depends on oil from a region that has suffered a catastrophe to provide it. Ecuadoreans are not amused; the pipelines are used as billboards for graffiti of the anti-imperialist sort, such as "The oil belongs to the people."

These two pipelines are akin to aortas connected to a network of steel veins that move oil from the wells and processing stations spread over the humid flatlands. The smaller pipelines aren't underground or routed away from roads and people, as they would be in a richer, better-run country. They rest on rickety pylons one or two feet high and just a few feet—or sometimes inches—from the roads. If you swerve into one of these pipes to avoid a pothole or lose control of your vehicle because you are drunk, you will create an oil spill. (It happens all the time.) Collisions are not necessary to create spills, because the pipelines are old and poorly maintained; they leak constantly. Until the 1990s, local residents say the environmental carelessness even involved the spraying of oil on dirt roads, so as to suppress the waves of dust that rose from them.

In the beginning, there was a misbegotten pioneering ethos. A new world was being carved out of the jungle; this was what progress was supposed to be. Hundreds of miles of new roads perforated the jungle, along with wells, waste pits, pipelines and processing stations. Boomtowns sprang up to house and entertain the thousands of construction and oil workers, and these insta-towns had the typical frontier accessories, including seedy bars, prostitutes and violence of the drunken sort. In a nod to the corporate creator of all these things, the regional capital was called Lago Agrio, which was the Spanish translation of the name of the American town where Texaco was born: Sour Lake. The transformation was accelerated by settlers who followed the oil roads. It is an irony of extractive industries that some of the damage they trigger is caused not by their own pollution but by settlement that is the result of opening up once-remote areas. "When a road is built, the settlers immediately come," a rain-forest activist told me. "And when they arrive, there is no control on the cutting of trees. That's how deforestation happens." Ecuador's government encouraged this process. To prevent Colombia from invading the mineral-rich region and to relieve overpopulation elsewhere in Ecuador, the government gave land to anyone who cleared the jungle and started farming.

During extraction, water is pumped into oil fields to force out the crude, and when the oil comes up, so does "produced water"—a constant burp of oil, salt and metals that can include benzene, chromium 6 and mercury. In less spry fields, as much as 90 percent of the liquid that comes out of the ground can be produced water, not oil. Instead of reinjecting the tainted water into the reservoirs or filtering out contaminants—standard practices now, and done in other countries back then as well—Texaco dumped the brew into unlined waste pits or poured it directly into the Amazon's rivers, according to environmental activists who are suing Texaco. More than 18 billion gallons of wastewater were disposed of in this way, they say, as well as 16 million gallons of oil—far more than the *Exxon Valdez* supertanker spilled into Prince William Sound. Texaco was lucky, because if you tip oil into Alaska's waters, everyone knows about it and cares deeply. In the Amazon in the 1970s and 1980s, not so much.

In Ecuador, Texaco burned off at the surface the natural gas that accompanied the oil when it came out of the ground, just as Shell did in the Niger Delta. As I mentioned earlier, if the amounts are large, as they were in the Oriente, they can be deadly for the environment as well as for the people who live nearby. In the Oriente, the burning of natural gas was so unrestrained and unmonitored that there is no reliable estimate of the amount of toxins released into the air.

It may not be much of an exaggeration to say that the rain forest was Texaco's rubbish bin. But that is not the worst part of the story.

In Lago Agrio I met Donald Moncayo, a leader of the Amazon Defense Front (also known as the Frente). Like any citizens' group whose name includes the words "Front" and "Defense," it was a no-frills outfit that operated out of a few chaotic rooms in a two-story building that had a warning sign at its entrance: "Site Under Surveillance." In case a visitor wasn't sure who was doing the surveilling, the sign helpfully showed an army tank. The Frente was involved in a multibillion-dollar lawsuit against Chevron (which completed a merger with Texaco in 2001) and counted among its adversaries not just the American oil giant but the Ecuadorean military, which at the time of my visit had lucrative security contracts with Chevron and other oil companies.

Moncayo, who was in his early thirties, had been raised in Lago Agrio because his father had gone there to find a better life. As a boy, Moncayo swam in a river that had veins of oil; the locals were unaware of the health hazards of the dark goo that was showing up in their previously pristine waterways. Moncayo worked briefly in the oil industry, which was the largest employer in Lago Agrio, but soon left for the ranks of environmental activists. It was not a lucrative choice—Moncayo now lived in a wooden shack with no running water or electricity, and when I stopped by one day, a very large pig was napping by his front door.

It was time for my Toxic Tour—this is the term Moncayo uses to describe the inspections he arranges for journalists. You could call Moncayo my toxic guide. We bounced in my rented 4x4 over the red dirt roads that crisscross the feeble jungles around Lago Agrio, trundling past rickety homesteads made of salvaged wooden slats. Most of these shacks did not have water, electricity, glass windows or mosquito nets; they were glorified lean-tos in which people lived their entire lives, deriving no benefit and, in fact, much misery from the oil a few thousand feet under the earthen floors of their homes. They looked impassively at me, another gringo who would come and go.

Our first stop was a dirt field the size of a soccer pitch that had, at its center, a rusted wellhead. Moncayo said the well had been shut down in the early 1990s. No problem so far. He then led me to a pond fifty yards away that was filled with murky water in which I saw globs of oil. Oil that had leaked into the ground long ago, when the well was in operation, was finding its way back to the surface. Dirt had recently been thrown into the water to absorb the oil, but this was the work of Sisyphus, because it would never end. "Every so often, more oil comes out, and they put more dirt on it," Moncayo explained.

We drove on for fifteen minutes, then stopped along a dirt road, where a farmer, Francisco Jiménez, told me an oil tanker truck had jackknifed into a stream adjacent to the road. Because the tanker was too heavy to be pulled out, Jiménez said, the oil had been dumped out of the tanker and into the stream. Jiménez walked to the stream and stuck a shovel into the ground, pulling out a clump of wet earth. The soil was streaked with oil, and the stream, I noticed, was topped with a film of crude. Jiménez told me that Texaco workers came a few days after the tanker accident and put some of the coagulating oil into sandbags, but most of it was untouched, seeping deeper into the ground and into the streambed. The oil-soaked sandbags remained where they'd been filled, just a few yards from the water. The company's workers could not be bothered to take them away, he said.

The tanker accident, Jiménez added, happened twenty years ago. The sandbags had been there for twenty years. Yet the pollution was still rising to the surface and was still alive in its unfortunate way. (Because the spill happened long ago and was minor in the grand scheme of pollution in the region, it was not possible to find a Texaco official who could deny or confirm it.) This was a reminder of one of the strange properties of oil: we may hastily bury it in the ground, but it does not disappear. Two things can happen. It may sink deeper, poisoning the groundwater, or it may rise to the surface, poisoning the water there. Or it may do both. We may wish to forget about oil, but oil will not let us.

To reach our last stop we drove down a dirt road that ended in a sickly, faded-green jungle; something was off. We walked a few hundred yards, pushing through the undergrowth, until we encountered what looked like a small, dark lake. It was actually a large pit filled with oil sludge that had the consistency of cookie batter. Where had the oil come from? It was dumped long ago, Moncayo said. Instead of disposing of spilled oil in an ecologically responsible way, Texaco would save money by pumping it into a tanker truck and driving the truck into the jungle, he claimed. With no one looking, the oil would be poured onto the ground, out of sight and out of mind.

"What would happen in Texas if there was a spill like that?" Moncayo asked, pointing at the black lake in front of us.

I said it would be cleaned up, quickly.

"We've been waiting seventeen years," he replied.

If a curious mind wanted to know what a major oil company might do without oversight in the developing world, the answer was to be found in the Oriente. And a remedy, in the form of a multibillion-dollar lawsuit pushed forward by a Harvard Law School classmate of Barack Obama's, could also be found there.

When Texaco discovered oil in the Oriente, Ecuador was a deeply impoverished country with a tiny industrial sector and no expertise in oil. It is only a small exaggeration to say that few Ecuadoreans, even the educated elite in Quito, could tell one end of a drill bit from the other. In the Oriente, Texaco's representatives were dealing with illiterate Indians who did not even speak Spanish, the country's official language. The first offering from Texaco, in exchange for permission from the Indians to look for oil, was a delivery of bread, cheese, spoons and plates. (The Indians threw out the cheese because it smelled so peculiar.) As in Nigeria and other underdeveloped countries, the foreign oilmen were met by an unsophisticated population that trusted their promises of good things to come.

Texaco eventually negotiated a contract that was typical in those days. It entered into a joint venture with a state-owned company that had been formed for the occasion; because Ecuador did not have a state oil company, one had to be created. Though it was a partnership, Texaco was the "operator," meaning it managed all activities. Because Ecuador's government had almost no expertise in the oil industry, it rarely questioned anything Texaco did. If Texaco said there was no problem dumping produced water into the Amazon, the government went along with it. In the early years of the venture, other than vague rules about respecting the health of the land and the people, Ecuador did not even have environmental laws for the oil sector. René Vargas Pazzos, a former army officer who presided over the state oil company in the early 1970s and was later minister of natural resources, said in an affidavit that Texaco had "complete autonomy" because the Ecuadoreans involved in the industry and in its oversight were either clueless or powerless.

"All of the members of the government of Ecuador assumed that the technology employed by Texaco was first-rate technology," Pazzos stated. "No one in the government thought that this petroleum giant utilized second-rate technology in Ecuador, including the dumping of production waters and other contaminants in the environment. No one ever questioned the practices of Texaco, for the simple reason that during all this time [no one] had the necessary information to question and oppose the practices which were chosen by an American company that presumably operated in an ethical manner around the world."

By the 1980s, a gathering disaster had fully gathered. Not only had the land in the Oriente been spoiled; the country's finances were ruined. The government had taken out billions of dollars of loans, on the assumption that it could pay them off with oil revenues, but when prices began to fall and a series of natural disasters struck the country, there was not enough money to go around. Thanks to oil, Ecuador had drilled itself into more than $10 billion of debt. The honeymoon with Texaco was over. The only good news was that Ecuador had slowly accumulated a moderate level of industrial expertise because some members of the new generation, sent to America for training, had returned home with engineering and other degrees. The state oil firm, a shell at its creation, was now capable of running the wells and pipelines that Texaco built. In 1992, as its joint-venture accord expired, Texaco left the country and handed over its facilities to Petroecuador.

This was not an entirely happy ending, because Petroecuador was not much better than its American godfather. The company cut corners as much as Texaco had, for the simple reason that the government, deeply in debt, needed every penny it could squeeze from its oil company. Instead of investing in better technology and safer practices, the government used oil revenues to pay off foreign debts and fund usual government operations (though in 1999 and 2000 the government defaulted on debt payments). Although Petroecuador has made some improvements in recent years, in the 1990s natural gas continued to be flared into the air, and polluted produced water continued to be dumped into rivers and leaky waste pits. The Amazon, victimized by a foreign corporation, fared little better at the hands of its state-owned successor.

Texaco had gotten out of the country but was not off the hook.

Surrounded by oil, Steve Donziger was fired up.

He stood beside a waste pit the size of a football field containing a dark liquid that looked like melted chocolate. Toxic rather than sweet, this was drilling's afterbirth, the hazardous brew of produced water. Because the pit's bottom was not lined with concrete or plastic, the black goo seeped into the earth and the aquifer that was just ten or twenty feet below. The air was being poisoned too, as an adjacent flare spewed forth its noxious fumes.

"If you don't have a headache yet, you will soon," Donziger shouted, and ten minutes later, after breathing the assorted chemicals and gases around us, I did.

Donziger is six foot four, handsome and blockheaded at the same time, like a hybrid of George Clooney and an Easter Island statue. With a Harvard law degree and a passion for justice of the anti-corporate variety, he was pursuing the huge lawsuit against Chevron for what environmentalists regard as the oil world's Chernobyl. The pit we stood beside had been carved into the rain forest by Texaco more than twenty years earlier. A few dozen yards away was another pit, half its size, also leaching poisons into the ground. And about thirty yards away from that, a swamp was filled not with swamp water but with oil sludge, which had leaked out or been dumped there. At the core of the lawsuit was an accusation that Chevron had created nearly a thousand sores like these, part of the shoddy network of wells and pipelines that had ruined the Oriente.

Donziger's mood, like the overcast equatorial sky above us, was dark.

"Companies cannot kill people," he said. "They cannot harm people. All of the deaths and suffering we're seeing as a result of this pollution was totally foreseeable by them. That's why they didn't do this in the United States. They thought they could get away with it here."

Good lawyers are practiced performers, expressing outrage on cue (and Chevron certainly denies Donziger's version). Donziger possessed these qualities; a former journalist who covered the dirty wars in Central America in the 1980s, he knew the utility of good quotes and the importance of good visuals. That's why we'd visited this petrowasteland and why he said, "This pit is a gun that is aimed at the lives of people. This is a toxic Three Mile Island." He was also a true believer. The case began as a class-action suit filed in a Texas court in 1993, and though Donziger was not involved at the very beginning, he soon joined the case and devoted his life to fending off Chevron's attempts to get the suit dismissed.

It was a legal suicide mission. If you want to sue oil companies, you need to be patient and fatalistic, because you are unlikely to get to trial; even if you do, after years of pretrial maneuvering, you are likely to lose; and if you happen to win, it will likely require years more to receive court-ordered damages because oil companies can afford to appeal and appeal and appeal. A hundred years ago, when John D. Rockefeller, the founder of Standard Oil, was told that a judge had fined his Indiana subsidiary $29 million, he replied that the judge "will be dead a long time before this fine is paid." Rockefeller then returned to his game of golf. In 1994, Exxon was on the losing side of an Alaskan jury's $5 billion verdict for punitive damages in the *Exxon Valdez* spill. Its appeals delayed payment until 2008, when the Supreme Court ruled that Exxon had to pay about $500 million; by that time, nearly one in five of the plaintiffs affected by the spill had already died.

Donziger, whether a missionary or a masochist or both, had persevered. In the decade after he received his law degree, he probably earned less money than anyone else in his Harvard class, including his basketball-playing buddy Barack Obama. That wouldn't surprise his fellow alums, because Donziger had spent his law school years organizing sit-ins of the dean's office to promote a course in public service law. After graduation, he had eschewed corporate jobs to work as a public defender and had written a book about inequities in the criminal justice system. He'd then taken on the mother of lost causes in writing a report about voting irregularities that had contributed to Al Gore's defeat in 2000. Since the 1990s, he had shepherded an apparently losing case to its likely demise—until fate, luck, nationalism and globalization intervened to create a huge surprise.

When a U.S. appeals court dismissed the case in 2002, saying it should be heard in Ecuador, Donziger refiled the suit in Lago Agrio. Surprisingly, the suit was accepted, and the trial began within months. "Chevron is in a situation they never imagined," Donziger told me in 2005. "They are a defendant in a court in a rain forest, getting their asses kicked. The trial they worked so hard to avoid is happening."

The case would seem easy to prove, with the billions of gallons of waste akin to blood on the still-slippery floor of a vast crime scene. To find proof, all you needed to do was stick a shovel in the earth, taste the tainted water that came out of the ground or inhale a lungful of the polluted air, as I did. You could visit the towns and see babies with deformities and people dying of cancer. It is because of the oil, you would hear. How could Chevron defend itself against a nation of evidence? Even the body politic had been poisoned. Thanks in part to the corrupting and debt-*accumulating* effects of oil, Ecuador has suffered chronic political instability; the country has gone through leaders like a shopaholic goes through dresses.

Chevron's lawyers were not paid to be stupid. They admitted that the Oriente was a disaster zone. But they blamed the mess not on the decades of exploration and extraction by Texaco, which spent $40 million on a belated cleanup program in the mid-1990s, but on Petroecuador, which had been running the show for a much briefer period of time. The pollution I saw, the shoddy pipelines and waste pits, the privations suffered by the people and the environment—Chevron said these were not its responsibility.

Ludicrous? The disaster certainly continued after Petroecuador took over but, as Donziger was quick to note, that's a separate issue. He's suing about the disaster that allegedly occurred on Texaco's watch. The $40 million remediation Texaco paid for after its departure was not only ineffective, it was a fraud, Donziger argued, noting that Ecuador's government was investigating whether polluted ground had merely been covered with fresh dirt and whether toxic waste had just been dumped elsewhere. (Chevron disputes these accusations, but in 2008, two Chevron lawyers were indicted in Ecuador on charges of fraud for their roles in certifying the cleanup.) I found it hard to imagine how $40 million, even if there was no fraud, could begin to address the environmental mess created before the 1990s.

There was another twist. How do you prove that contaminants in the soil were left by Texaco before it pulled out in 1992 rather than Petroecuador afterward? Thousands of soil samples had been taken for the trial, but they had not been tested for age. The results showed the amount of pollutants but not how long they'd been in the soil. And the samples—taken by teams working for Chevron, Donziger and the court—were not entirely consistent. The preponderance showed high levels of pollution, but some did not. It was mind-numbing stuff. Scientific and legal documents submitted in the case surpassed 200,000 pages, which is more than the judge, who had just one clerk, could be expected to digest.

That was why Donziger's political skills were as crucial as his legal touch.

"This case has to be won both in and out of the courtroom," he told me. What he meant—and this was a lesson to other lawyers who might sue extractive companies—was that getting the right judgment and getting a payment would require more than good arguments in court. The political winds needed to blow, or be made to blow, in the right direction.

On this, Donziger got billion-dollar lucky.

Justice is rarely blind in Ecuador. It is rarely blind anywhere. Until a few years ago, verdicts in South America did not favor the little guy taking on big corporations that had lucrative ties to the ruling class. But in 2006 Ecuadoreans elected a leftist president, Rafael Correa, who joined Hugo Chávez and other soul mates ruling in Bolivia, Nicaragua, Brazil, Argentina, Peru and Chile. In Venezuela, Chávez famously forced oil companies to rewrite their joint-venture contracts, while Bolivia's president, Evo Morales, also moved to nationalize his country's oil and gas industry. Even before Correa was elected in Ecuador, the government had imposed a windfall tax on oil companies and terminated a contract with Occidental Petroleum. Courts and judges, rather than protecting Big Oil, had become weapons against them.

In 2007, the presiding judge of Donziger's case asked for a remediation and restitution estimate from an independent expert; the expert's report put the cost, including payment for cancer deaths, at up to $27 billion. The hammer, in the form of a staggering verdict, seemed ready to come down on Chevron. But a favorable ruling would get Donziger only half the way home. Chevron can appeal in Ecuador and international courts to delay paying a dime until Donziger, who is in his forties, qualifies for Social Security. Remember, fourteen years elapsed between the verdict against Exxon in Alaska and actual payment of punitive damages by the company. How might Donziger—how might any public interest lawyer—get compensation in a timely manner?

"We could appeal to their conscience," Donziger said. "We could say, Come on, guys, you can afford this. But they don't function that way. Companies have a fiduciary duty to sharcholders to not pay stuff that they don't have to pay. At the end of the day, it's a political struggle to enforce them to pay."

It didn't take long to see what he meant.

I arrived in Lago Agrio a day before the judge was scheduled to inspect the Guanta processing facility, built by Texaco in the 1970s. The heart of the trial consisted of inspections of toxic sites of this sort, and the inspections had a circus feel. The judge arrived with his experts; Chevron's lawyers arrived with their experts and a few bodyguards; Donziger's team showed up with an ensemble that usually included plaintiffs in the class-action suit; and often there was a handful (or more) of journalists. Samples were taken, occasionally with arguments over precisely where the samples should be taken from, because contamination is not evenly distributed at polluted sites. Testimony was heard in the shadow of flares and wells.

The night before the Guanta inspection, Judge Efraín Novillo abruptly announced a postponement. This was a disappointment to Donziger, because the Frente had brought a busload of journalists to town from Quito. (I was not part of the group.) With no inspection, the journalists would have no anti-Chevron material to write about.

Donziger's father was a businessman and his mother a social activist, so he is an entrepreneur of indignation; he knew to counterattack.

In the morning, Donziger led a few dozen indigenous Indians and environmental activists to the town hall, a four-story, tinted-glass structure that looked as though it had not been washed since it was built, decades before. It projected neglect. Judge Novillo's chambers were on an upper floor in a small, book-lined room that had never before been invaded by a posse of lawyers, reporters, students and Indians. A book by Gabriel García Márquez lay on the judge's desk, and he now faced a real-life version of magical realism as his uninvited guests demanded to know why he'd called off the inspection.

"Texaco organized this!" shouted a law student from Quito who wore a bandanna around his neck in the imagined style of Che Guevara. His class was on a field trip with their professor, who was a member of Donziger's legal team.

Judge Novillo, a quiet man who looked on the tired side of his fifties, reacted calmly.

"This order is from state security," he replied, briefly displaying an official-looking piece of paper.

"Let us see it," someone urged.

"I cannot show it," Novillo said.

Howls of outrage filled the judge's chamber. To quiet things, the judge started to read the order, which he said had come from the local special forces base, Rayo-24.

"'Military intelligence fears there could be a hostile situation,'" he began.

Another interruption.

"You are not showing it!" the law student shouted.

The judge flashed the letter again, but not for long enough.

"This is a manipulation from Texaco!" someone yelled.

"We have been hurt by the pollution," a short Indian woman wailed.

I heard a baby cry. With all the bodies packed into the room, the humidity had reached the saturation point. For the judge, there was no escape. He was forced into a sort of data striptease, with information

Lieutenant Colonel Francisco Narvaez confronts indigenous protesters and journalists outside a military base in Lago Agrio.

revealed piece by piece to the aroused crowd. Finally, he held out the letter for everyone to read. It seemed right that the letter become public, but the process was not pretty, because the judge had been bullied. It took little imagination to guess what might happen to him if his eventual verdict did not satisfy the plaintiffs.

It was time for the next act. The buses of Indians and lawyers and students and journalists rolled across town to a military base. Everyone stood outside the gate in a light drizzle, armed with cheap umbrellas. The base commander, Lieutenant Colonel Francisco Narvaez, soon emerged, wearing fatigues and a burgundy beret.

"I do not know about the existence of the letter," Narvaez told the crowd. His statement was met with disbelief, because the letter was signed by his second-in-command. It was unlikely that a letter of that sort would be issued behind his back.

"I am going to investigate it, and when I find the answer I will meet with you and tell you what is going on," Narvaez added.

Donziger, sodden but spirited, decided to try something new.

"The Texaco staff is staying here," he told Narvaez. "There is an agreement."

This was cast as an accusation.

"There might be," Narvaez responded uneasily.

Donziger had known for months that Chevron had built a villa at the base and had agreed to give it to the military once the case ended. Donziger hadn't opposed the deal because Chevron was not popular in Lago Agrio; he'd realized that the company's lawyers would be safer with military protection. But with more than a dozen news-hungry journalists recording the moment, Donziger suspected that the time was right to accuse the military of being on the payroll of gringo oil-men. He was correct. The accusation made national headlines, and a little more than a month later the Ecuadorean military canceled all military contracts with oil firms and ordered Chevron off the base. The military was also forced to publish the now-discredited contracts, and this was yet another political victory for Donziger, because it showed that Chevron had contaminated the army, too.

Donziger was winning his battle, but it was about the past. What remained of the Amazon—and there was much in Ecuador that was untouched—was in jeopardy. Oil had become too valuable to be left underground, so foreign oil companies were seeking permission to drill in the southern portion of Ecuador's Amazon. There, another type of activist stood in the way, armed not with a law degree but with a spear.

Driving south from Lago Agrio I passed through a fifty-mile stretch of apocalypse—a mutant panorama of oil fields and gas flares in which crude oozed and burned around me. Oil was also in the melting asphalt under my car, and it coursed through the leaky pipes that curved felinely along the road, inches from my fender. I drove through the town of Coca, where an anemic jungle gave way to the offices of Halliburton, its front gate painted in the company's by now familiar red and white. I continued farther south, the road turning to dirt as it ascended into cloud forests before descending into Puyo, the gateway to Ecuador's southern Amazon.

I navigated through Puyo's narrow streets to a small house on a dirt road in a poor neighborhood where the windows were covered with iron grilles. This was the beyond-modest political office of the people of Sarayaku, an indigenous tribe that lived fifty miles away in the Amazon and was trying to prevent oil drilling on its territory. Inside the house, the tribe's president, Marlon Santi, was talking on a shortwave transmitter to his friends in the jungle. When he was done, we went out for pizza.

Santi, whose impeccable mane of dark hair was tied in a ponytail that fell down his back, was everything the Indians in the Oriente had not been when Texaco arrived in the 1960s. Santi was, to begin with, educated. He'd attended a Catholic high school in Puyo, so he spoke and read Spanish flawlessly. He had visited the Oriente and knew what had happened there. Thanks to groups, such as Amazon Watch, that facilitate foreign travel for indigenous leaders, Santi had visited other countries coping with similar extraction issues, and he'd even visited the United States, where he'd met Chevron shareholders and had spoken to them about the threats to his Amazonian homeland. He'd learned the most important lesson of all—that it is insufficient to just say "no" to oil companies. The rain forest may be a brief paradise for ecotourists who stay two nights and sleep under finely woven mosquito nets, but living there in primitive conditions is tantamount to a lifelong sentence of malaria and malnutrition. Santi's "no" to extraction must be accompanied by a "yes" to his people's desire for medicine, education and other virtues of modernity. It is the same in other developing nations— impoverished communities cannot be expected to turn down the seductive promises of extractive development unless there is an alternative.

The next morning, to learn more about Santi's unconventional war, we flew into the Amazon in a single-engine Cessna.

The Americans who found oil in the Oriente in the 1960s were actually the *second* wave of petroleum geologists in Ecuador. The first wave, led by a company called Leonard Exploration, arrived in the 1920s and looked for oil in the south. To move their equipment into the frontier region, the company built a highway to Puyo, which at the time was a tiny settlement on the lip of the southern Amazon. Leonard found no oil and was succeeded more than a decade later by Shell Oil, which built an airstrip outside Puyo. Shell also failed to find oil and pulled out in the 1950s, but its mark remains; the Cessna I squeezed into took off from a one-runway airport named Shell.

After just a minute or two in the air, Puyo was behind us. As we flew south, beneath us and around us was the boldest rain forest I had ever seen. This was untouched Amazon, the trees thick and luscious, squeezed together like broccoli stalks to the horizon and beyond. It was a visual incarnation of nourishment, the vegetation forming a blanket of delectable, living green.

"From here, it's our territory," Santi yelled over the whining of the engine, gesturing toward a shelf of hills.

What he meant when he said "our territory" was Sarayaku, where modern geologists, with better technology than their ancestors, believe there are vast reservoirs of oil. The old question of who owns it returned to the forefront. According to Ecuadorean law, the indigenous people own the land but the government owns the minerals underneath it. It's a built-in conflict, because the government needs to use the land to get at the oil. In 1996, the government awarded an exploration concession to an Argentinean company, and after several years of fruitless negotiations with the area's residents, most of whom were not seduced by offers of cash, the company and the government unilaterally went ahead with seismic testing in 2002.

This was a mistake.

Santi pointed to a river I could barely make out through the dense trees.

"This is where the company tried to set up their headquarters," he shouted. "Our resistance started there."

Santi called it a war, but no shots were fired. The oil company used helicopters to drop a team of geologists into a clearing in the jungle. They planned to conduct seismic tests by setting off small explosions underground. Sensors would monitor the shock waves, and the data would help make a map of the underground geology. This is a prelude to drilling. The Ecuadorean military provided a security detail—a handful of soldiers who were dropped into the jungle.

As large as the Sarayaku territory is, outsiders cannot sneak in by helicopter and set up a base camp without someone seeing or hearing the activity. And the people of Sarayaku were ready—they knew of the company's plans and had begun patrols along their territory's borders.

Within hours of their arrival by helicopter, the soldiers and oil workers were surrounded by spear-carrying Sarayacans whose faces had been daubed with black war paint. The soldiers surrendered without a shot being fired or a spear thrown. They were taken to Sarayaku's main village. After several days of negotiations, they were released in exchange for a government promise to never let an oil company enter the territory without explicit permission from its people.

This part of the war never ends. Oil firms are not like door-to-door salesmen who, turned away from one house, go to other houses, other streets, other towns. There are a finite number of reservoirs in the world, so oil companies have a limited number of doors to knock on. They keep knocking even after leaders tell them no. It is not an exercise in futility. The campaign for drilling continues because oil companies know that local and national leaders can change or be changed. It is just a matter of finding the right price and offering it at the right time to the right person.

After thirty minutes of flight, Santi's alternative answer came into view. The Cessna circled over a clearing of thatched huts and dropped to a bumpy landing on a dirt airstrip. I had arrived in Sarayaku, and after unloading my backpack and standing clear as the plane turned around and hopped back into the sky, I was wrapped in the thick heat and vibrant noise of the Amazon.

It was a mile-long hike to the village center. Santi and I walked past small huts built under the canopy of trees, invisible from above. We crossed a shaky footbridge suspended over a furious river, climbed a set of steps carved into a hillside and emerged at the center of Sarayaku. Initially, it was not much to look at—about a dozen huts of different sizes, grouped in an uneven circle around a dirt clearing the size of several football fields. After dropping my bag under the shelter of a hut that had a thatch roof but no walls, I went for a tour with Santi.

Behind the main circle was a school composed of wooden buildings; the instructors were several volunteers from Spain who also ran a small infirmary. The most stunning sight was a row of solar cells. I noticed that one of the shacks contained several computers. In isolated rain-forest communities, such attributes of modernity are rare. There was even an open-air hut that served as an entertainment center of sorts—it held a television and VCR that ran off solar power. Inside, a dozen German students listened to a Sarayaku leader wearing a "Viva Zapata" T-shirt describe the community's battle against oil companies. I listened too, swatting at mosquitoes. "Even if they give us one million dollars, we don't want it," he told the Germans. "Thirty years of oil has not benefited Ecuador. The oil areas have pollution, disease, narcotraffickers and violence."

Revenues from tourists are a lifeline for Sarayaku. Visitors don't pay much—the accommodations consist of almost-bare huts that don't even have mosquito nets, and the nearest chilled drink is more than fifty miles away, in Puyo. But ecotourism constitutes a stream of alternative revenue that makes it possible for Sarayaku to say no to oil extraction. When the village youths got a soccer game going, I noticed that most of them had sports shoes and jerseys with numbers on their backs. In a rain-forest community, these were totems of prosperity.

The noonday heat crushed us. Santi, dressed in a black T-shirt and blue-jean shorts, led me back to the hut where I had dropped my backpack. We lay on the floor because it was a strain to sit up as the humidity and heat reached headache levels. Suddenly a young man rushed past with a bandaged hand—he had just been bitten by a snake and needed an antivenin shot that might be available in the infirmary. If not, he would die, because the eight-hour canoe ride to Puyo would take too long and there were no more Cessnas today. This was life in the jungle, brutal even with the totems of modernity Sarayaku possessed.

As oil prices rise, so do the incentives from companies wishing to drill. Already, Santi told me, oil money was dividing Sarayaku. The leader of a subtribe that wanted to cooperate with the companies had been accused of treason and forced into exile with his family. A pro-oil tribe that lives between Sarayaku and Puyo refused to let Sarayaku's obstinate leaders use its rivers for passage to the city. Santi had to ply longer routes or cadge rides on the occasional Cessnas. Even though not a barrel of oil had been brought to the surface in Sarayaku territory, the liquid was causing trouble. All of Ecuador was contaminated by oil.

In 1961, Stanley Milgram conducted one of the most famous behavioral experiments of all time. To measure people's tendencies to obey authority, Milgram, a social psychologist, asked his human subjects to deliver electric shocks to a person who failed to answer questions correctly. The person receiving the shocks was in fact pretending to be in pain—there were no shocks—but the subjects didn't know that. Despite increasingly loud screams and pleas from the pretenders, most of Milgram's subjects complied, administering shocks that seemed to cause extreme pain to complete strangers. "Ordinary people," Milgram concluded, "can become agents in a terrible destructive process. Moreover, even when the destructive effects of their work become patently clear, and they are asked to carry out actions incompatible with fundamental standards of morality, relatively few people have the resources needed to resist authority."

Milgram's findings have been used to explain a wide range of otherwise perplexing human activities, from the willingness of ordinary Germans to participate in genocide during World War II to the readiness of Enron traders to create energy shortages that caused a painful spike in electricity prices for ordinary people. (In a taped conversation, one Enron trader said admiringly of another, "He steals money from California.") I had Milgram in mind while interviewing oil executives across the globe. I was interested not only in the relatively few who'd been indicted or convicted of corruption but also in the larger number

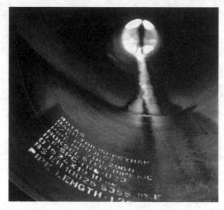

A section of the Baku-Ceyhan oil pipeline

who woke up every morning, ate a good breakfast, kissed their wives or girlfriends good-bye and headed out to participate in an extractive industry that had a high probability of bringing poverty, violence and dictatorship to the countries they worked in. Milgram provided part of the answer, as did Frank Ruddy.

If you walked past Frank Ruddy on a street in Washington, D.C., you wouldn't look twice. Just another silver-haired lawyer with framed degrees on his office walls, you'd think. Ruddy graduated from College of the Holy Cross in 1959, then got a law degree from NYU and a PhD from Cambridge University. From 1974 to 1981 he was a top lawyer in Exxon's legal department, and after that he joined the Reagan administration, becoming ambassador to Equatorial Guinea. Ruddy was a loyal insider of both the oil industry and the American government. Then he got whacked in the side of the head by a thing called reality.

In most countries, an ambassador is insulated from day-to-day life by a large staff that caters to his needs, by unenlightening contacts with government officials who tell him sweet nothings, and by the expectation that he will hew to whatever line the State Department has established. The opportunity and need for independent thought are not great. But in Equatorial Guinea Ruddy headed the tiniest of embassies, the ministers he met were thinly disguised thugs who could barely read (some were thought to be functionally illiterate), and the country was considered so unworthy of a superpower's attention (this was before oil was discovered) that nobody in Washington cared what Ruddy did or said. He was on his own.

After serving in Equatorial Guinea for four years he returned to Washington to become general counsel in the Department of Energy, and when the time came to move on from that job, he decided to practice law on his own. He was on track for a snoozy finish to a quiet career when Equatorial Guinea reentered his life. After oil was found, Ruddy realized from afar that Equatorial Guinea had become a worse hell house than before, because Obiang's theft and repression had only increased. Initially, he watched in quiet dismay as Exxon and other companies, supported by the U.S. government, filled Obiang's pockets with hundreds of millions of dollars in oil revenues. Then he began to speak out publicly, becoming one of the regime's harshest critics as well as a skeptic about the corporate way of dealing with the world.

"I am not anti-capitalist, but capitalism can be its own worst enemy," Ruddy told me when I visited the two-partner law firm that was his obscure (by Washington standards) place of business in 2004. His no-frills office was Dilbertian in its lack of pretense or luxury, with all the charm of a medical waiting room. "It causes things that are horrible and threaten to undermine it, like exploiting the poor. I know some people well in Exxon, and they will say, What can I do? I am a lawyer seventeen layers from the top. Others will say, It's not our fault. We didn't elect these sons of bitches. We're just going to make money and get out."

These excuses are truths, too. A midlevel lawyer or executive cannot change the priorities of a multinational whose survival depends on access to new oil fields. A corporation is akin to the Milgram authority figure who wears a white lab coat and calmly tells you to turn a knob. As a sweetener, money can be made by doing as you are told. Unlike Milgram's subjects, who heard and saw the consequences of their actions, most executives do not come face-to-face with the poverty and violence their industry can foment; the closest most of them get to subsistence misery is from the inside of an air-conditioned sedan that takes them from one meeting to another in a foreign city. They could connect the dots if they wished, but it's easy and profitable not to do so.

"These aren't stupid people," Ruddy continued. "Evil is not people with mustaches who look like they're doing bad things. Evil is done by people in suits sitting in boardrooms making horrible decisions. They do it because it's worth it."

Alain wore an impeccable suit, with a silk handkerchief tucked into his breast pocket in a dapper, just-so way. He was French, spoke a handful of languages and carried himself in an elegant manner, as though he had just emerged from the Élysée Palace or the pages of *Madame Bovary*. Alain was a senior executive at a West European oil company, and his elegant, drawl-free manner was the opposite of the slouching-toward-Riyadh gait of other oilmen. Alain had won multibillion-dollar contracts in the crucibles of oil, from the Middle East to Africa, so he knew the industry and enjoyed being known as knowing it.

We met for breakfast at a luxury hotel in Manhattan and talked about warfare, because the U.S. invasion of Iraq was on the horizon. Alain, who had been to Iraq many times and gave the impression of knowing Saddam Hussein, accurately predicted that Iraqis, though glad to be freed of Saddam, would resist an occupation. But he evoked the likelihood of an anti-American insurgency in passing, as though it would just be the *drôle de guerre* before the fiercest combat of all began, the one for Iraq's oil.

For Alain, geology was nearly as sexy as sex itself. He excitedly talked up the physical wonders of Iraq, which has some of the largest reservoirs in the world, with high-quality oil close to the surface and easy to move to seaports. His awe was of the sort Frenchmen usually reserve for discussions about the beauty of the mistresses they possess or desire. Yet it was not just the loveliness of Iraq's fields that moved Alain's heart—it was the fact that exploration deals for fields of any promise, let alone great ones like those in Iraq, were rare and getting rarer.

"Iraq has everything," he said. "It is a must for oil companies to go into Iraq because access to crude is the essence of an oil company. Without oil, what is their purpose?"

Executives like Alain, responsible for winning exploration and production contracts, are the gladiators upon whom oil firms depend for survival. In his voice I heard the pride of a man who'd survived the all-or-nothing combat that takes place in the offices, palaces and bars that are oil's coliseums. He described the process as a "battle of giants" and a "fight to the death," because small firms are crushed by larger ones in today's world. As in ancient Rome, if you finish second you are dead. "It's a natural war, below the belt," Alain added with a smile.

He had just raised an issue I was curious about. With the stakes so high, is there anything an oil executive will not do?

Alain was a classic raconteur. A listener savored his words, despite trusting only half of them. He was evasive when discussing the shortcuts his cohorts were known to take. In recent years, executives from his firm had been convicted of bribe making, bribe taking and a multitude of other financial crimes related to illegal dealings in the Middle East, Africa and Europe itself. Some of Alain's friends were felons. But at the mention of corruption, he expressed the despair of the maligned.

"Even my family and friends think that when I go to Saudi Arabia, I am carrying suitcases of cash to bribe everyone," he said.

I didn't need to ask if his family was right. His expression, of the "How could they think that?" variety, invited no follow-ups on the contents of his Mideast valise. I took a shortcut of my own, mentioning that American executives had said the French were the most corrupt. His eruption was so raw that crumbs of the croissant he was eating began shooting across the table.

"That's bullshit. Do you think ExxonMobil acts like a saint in Nigeria? And Marathon in Gabon? You think they are saints? Bullshit. Look at the corruption scandals you have had, like Enron. Do you think these things do not happen in your oil companies, too? You have men like this in ExxonMobil right now."

He was so mad that he forgot to proclaim his innocence. After breakfast, one of his aides, who was British, escorted me to the street and listened as I innocently asked whether I should trust the French who said the Americans were the most corrupt or the Americans who said the French were the worst.

"You must check everything," she advised softly.

On an April morning in 2006, I visited the federal courthouse in Manhattan. After giving my cell phone and digital recorder to a security guard, I rode an elevator to the eleventh floor. There, sitting on a hallway bench, was a distinguished-looking man wearing a pin-striped suit with a white handkerchief tucked into the breast pocket. On his right were two lawyers from Cooley Godward Kronish, a firm that specialized in white-collar crime. Seated on his left were two more lawyers from Cooley Godward. The meter was running on this costly legal team, at least $2,000 an hour in all, but the well-dressed client could afford to wait.

I introduced myself. There was a brief silence.

"Are you looking into the relationship between the industry and the government?" the wealthy client asked.

"To the extent that I can," I replied.

"That's where to look," he said, then smiled.

James Giffen was the son of a California clothier and had come a long way from his childhood in Stockton. His ambition had propelled him to a career in the oil field industry, and this had sent him to the Soviet Union in the 1960s. Giffen did not speak Russian, but he was a schmoozer of the highest order, capable of ingratiating himself with powerful men by boasting persuasively about accomplishments and clout that did not always accord with reality. In the 1980s he even managed to befriend Mikhail Gorbachev, the Soviet leader who began a new era of openness with the United States, and put together a consortium of American firms that offered to make the largest investment in the Soviet Union since the time of Lenin. That deal fell through, largely because the Soviet Union itself was falling through, but Giffen did not disappear. He emerged as the top financial adviser to Nursultan Nazarbayev, leader of newly independent Kazakhstan.

In those days, American and European firms were brawling like roughnecks over contracts to explore Kazakhstan's reservoirs, including one of the world's largest fields, Tengiz. Developing these fields would cost scores of billions of dollars, and the payoff would be glorious—Kazakhstan's oil and gas could be worth more than $1 trillion. Giffen was custom-made for these above-the-law, fortune-making times. He was intelligent and aggressive as well as profane, hard-drinking and, when the situation required it, fierce. Giffen convinced Nazarbayev that foreign oil companies were predators—and this was certainly true, because the behemoths from Houston, London and Paris knew that a just-born country like Kazakhstan was an excellent locale for a financial killing. Giffen, who became so close to Nazarbayev that they shared saunas as well as confidences, was the regime's pin-striped bulldog.

As deals were struck, Giffen's wealth soared. Chevron, which in the Soviet era won a contract to develop Tengiz, had agreed to pay Giffen 7.5 cents for every barrel extracted—a "success fee" worth tens of millions of dollars. Giffen was delighted to show off his new wealth. In New York, he bought a posh estate in the suburb of Mamaroneck and drove into the city in an $80,000 Bentley. The walls of his office were decorated with pictures of himself and Gorbachev, President George H. W. Bush, President Carter, President Ford, President Nixon, President Clinton and so on. The glory days lasted until 2003, when Giffen was arrested at John F. Kennedy International Airport as he was departing for Kazakhstan.

According to the Justice Department's indictment, Giffen had diverted nearly $80 million in payments from oil companies and channeled the funds to President Nazarbayev and Nurlan Balgimbayev, a former prime minister. Most of the funds were routed through Swiss accounts. Some were conveyed in gifts, including $30,000 in fur coats, two snowmobiles and a luxury speedboat. Remarkably, the corruption indictment described President Nazarbayev not as a victim but as a partner in these alleged crimes. The Kazakh leader was named as an unindicted coconspirator in a scheme to steal the country's oil revenues.

I saw Giffen several times at the courthouse and visited him at his Manhattan office. (He had no trouble making his $10 million bail.) He was always impeccably dressed and reminded me, with his well-groomed hair and his prosperous look and abundant vitriol, of Lou Dobbs, the never-at-a-loss-for-outrage host on CNN. At his office, Giffen went on for hours and hours about deal making in the Caspian, scribbling charts on a legal pad to illustrate one of his points: that an honest man willing to make a fair deal is unknown in the oil world. A condition of our meeting was that I not quote him, and that's just as well, because profanities were among his closest verbal friends. His kindest description of oil executives was along the lines of "They want to fuck you against a wall." He suggested that I burn into my head the notion that oilmen will do anything to win. Giffen, who was honest enough not to exclude himself from their anything-goes ranks, help-fully directed my attention to a notorious scene in the movie *Syriana*. An oilman, played by Tim Blake Nelson, explains how the system works: "Corruption is our protection. Corruption keeps us safe and warm. Corruption is why you and I are prancing around in here instead of fighting over scraps of meat out in the street. Corruption is why we win." Giffen proudly reminded me that Nelson's character was based on him.

In pretrial hearings, Giffen's legal strategy was startling. He did not contest the financial actions of which he was accused. Instead, his lawyers asserted that he was, in addition to being the Kazakh leader's close friend, top adviser and partner in saunas, a conduit of information for the Central Intelligence Agency. And it was true—Giffen's contacts with the CIA and other U.S. government agencies were not disputed by the agencies or the prosecutors who filed charges against him. His lawyers were preparing what's known as a "public authority defense," which means that an accused's actions were committed with the knowl-edge of the U.S. government and thus cannot be prosecuted. Giffen was our man in Astana. Giffen's lawyers asserted that the U.S. govern-ment was in a position to know about the diversion of Kazakh revenues into the president's pockets.

By the middle of 2009, six years after indictment, the case still had not gone to trial. Due to the potential embarrassment of revealing more information about the CIA's role in Giffen's activities, the charges seemed likely to be dropped or plea-bargained to a slap on the wrist.

What is your number?

I mean, how much would it cost to corrupt you?

Particularly for oilmen, this is not a hypothetical question. The industry all but extracts cash from the ground. Making steel, computers or socks requires the purchase of raw materials, the building of factories, the employment of workers and a search for buyers. The oil industry cuts down on that: if you acquire ownership or rights to a field, you acquire the key to a vault of guaranteed profit. You don't need to skim wealth; you can scoop it up.

J. Bryan Williams, who was a Mobil executive in the 1990s, knew this as well as anyone. Making deals across the former Soviet Union, Williams encountered offers and demands for under-the-table payments. This was in an atmosphere of negative regulation—government officials who were supposed to enforce the law were instead inviting him to break it.

Oil rigs in the waters of Azerbaijan

We met on a winter day in 2006. Williams, who'd pleaded guilty to avoiding taxes on income that prosecutors said had come from illegal payoffs, had just been released from prison and was living in a halfway house in Washington, D.C. He was sharing a room with drug dealers. Permitted to leave during the day, he agreed to meet me at a middle-brow Asian restaurant near K Street, where the capital's lobbyists, including ones from his former employer, have their offices. Williams wore a rumpled plaid shirt and cotton pants that were similarly distressed. He had the thick, disheveled look of a retired lumberjack who'd just rolled out of bed. This was a change from his Big Oil days. Back then, he'd dined at the best restaurants and had flown first class when a company Gulfstream was unavailable. He'd put together billion-dollar deals. In addition to several years in jail, his sentence had included more than $3.5 million in fines and restitution, so I did not know whether his fallen-on-hard-times appearance reflected his reduced circumstances or was a charade until he was free and clear of the judicial system. Lunch was on me.

Williams began by saying what he had not done. Once, Russian businessmen who were in a position to help Mobil had asked for a $5 million "loan." Williams said he'd nearly fallen out of his seat because their request was so crude and unrealistic. Another time, an aide to a powerful minister had said a $100 million payment would be required before the minister would approve a particular project. Williams told me that he'd declined but had later heard that the minister received the tribute from a rival firm. And another time, during a legal dispute over Mobil's ownership of a refinery, a senior KGB official had offered, for a price, to have his organization make the problem go away. Williams had turned down the offer but asked, hypothetically, how the KGB would settle things. His KGB friend had replied, "We will occupy the factory."

Williams joined Mobil in the 1970s and initially worked on deals in Saudi Arabia and Nigeria, though by the 1990s he was deeply involved in Russia and Kazakhstan. He said French and medium-sized firms of any nationality were the worst bribers, because they did not have the mass or mastery of the truly big players. As he put it, "They have to do the extras." Mobil's tactics were legal, he said. Though he didn't want to get into specifics, he offered several scenarios that obviously came from his memory rather than imagination. In Nigeria, a tribal chief whose support was needed for a project might be hired at a high salary for which clocking in for work would not be expected. Depending on the salary, these men were referred to, in intracompany discussions, as "double chiefs" or "triple chiefs." Relatives of key officials might be hired under similar conditions.

As a convicted felon, Williams did not have an unblemished reputation to defend, so he spoke with greater honesty than most oilmen do.

"What are oil companies supposed to do? We don't create these places. Do we only develop oil in London and Paris? If so, we'll all be out there walking and stepping over piles of manure."

He was getting angry.

"You go where the bloody oil is," he said.

Like most people, Williams was not averse to becoming rich. Not Miami-condo rich, which he could afford on his midlevel executive salary, but mansion-in-Aspen rich, which would require a bit more. And why not go for it? Fortunes were being amassed by everyone he dealt with—ministers, bureaucrats, consultants. According to his plea agreement, in 1993 he set up a Swiss bank account into which was deposited more than $7 million. Prosecutors alleged that a portion of those funds were kickbacks, but Williams denied this. In court, he carefully admitted that "some of the payments . . . including a $2 million payment I received in 1996, were paid to me by people, organizations or governments with whom I did business on Mobil's behalf."

When I asked about the $2 million, Williams said it was a loan out of the blue by associates who were grateful for his hard work on a pipeline deal. He shifted in his seat, looked down at his plate and swallowed his words as well as his food. "I thought they were going to give me a watch," he mumbled.

He was more persuasive about the guilt of others, and this reminded me that finger-pointing can be a compass as well as a dodge. He mentioned Viktor Chernomyrdin, the former Russian prime minister reported to have accumulated a fortune of more than a billion dol-

lars during the grab-what-you-can 1990s. When the CIA sent the
White House a classified report that said Chernomyrdin was corrupt,
Vice President Al Gore, who had close ties to Chernomyrdin, reportedly returned his copy with a single word scrawled on the cover: "Bullshit."

"The governments are more crooked than some of the oil companies," Williams said.

I had several hours, on the train back to New York, to consider the
dissonance of a government imprisoning fish like Williams while protecting sharks like Chernomyrdin. The White House had rolled out a
red carpet for Azerbaijan's father-and-son despots, Heydar and Ilham
Aliyev. Nazarbayev, accused by prosecutors of receiving millions of
dollars in bribes from Giffen, was treated to a state dinner in Washington while his partner in kickbacks, Giffen, faced the prospect of the rest
of his life in jail. I had to wonder whether the sins of my bitter lunch
companion were the symptoms of an illness rather than the illness
itself.

Sadad al-Husseini, who served as vice president for exploration and
development at Saudi Aramco, was credited with turning the state-
owned company into a model of efficiency and safety in the 1980s and
1990s. After retiring from the company in 2003, he became a global
consultant, working out of a villa in Dhahran. He was not royalty-rich
but was wealthy enough to afford to live in another villa not far away.
He was one of the most respected oil experts in the world, which is why
I listened closely when he described the behavior of foreign executives.
He noted that Aramco had more oil at its disposal than any other company in the world and did not need to compete for new fields or worry
about hostile takeovers if its financial performance faltered; the firm
could afford to operate its facilities in a first-class manner. The situation was not the same at companies owned by investors rather than
royalty.

"Their motivation is totally different from our motivation," Husseini said, using a vocabulary that was as carefully selected as the numbers in an engineering calculation. "Some of my best friends are in

these oil companies. They have brilliant leaders, they have brilliant engineers, but they get exposed to commercial pressures which they have to live with. If they are in financial trouble and have to cut corners, they will cut corners. It means that if your tanker is old and you ought to retire it, you keep it working. It means that if you have an offshore platform that is beyond the national boundaries of a certain country and you can dump chemicals into the sea, you do. It means that if you have to abandon a facility that is a pollutant, you abandon it in place and walk away without cleaning it up. If you've hired people and can work them in unhealthy environments where you've got sulfur dioxide, you do it. All of these are ways in which you say, It's not my problem. It's not my cost."

I liked most executives I met. They were hardworking men with a thrill for the deal, a fear of failure and a moral compass that occasionally responded to the force of self-preservation; the compass did not always point in a moral direction. They were flawed, but that meant they were like the rest of us, using masks and shields in their real-world dealings. They might be one thing at the office, perhaps believing in the institution that employed them, perhaps not believing; and they might be something quite different outside the office. Gabriel Nguema, the son of one of the world's worst dictators, Teodoro Obiang, had a reasonable perspective on the oilmen who marched through his office at the Ministry of Mines, Industry and Energy. "Somebody in a company told me that when you work in this company you take your brain out and you put this box in," Nguema said. "Once you return to your house you put your brain back in. So you don't follow your feeling of what people need, you just follow what the company tells you." Nguema was on to something, though he didn't attempt a broader application of his observation. Oilmen had the same business DNA as executives in other industries. Buy low, sell high, keep your job. The main difference arose from the unusual conditions in which oilmen competed.

If you want to alter the behavior of an executive who usually follows the highest ethical standards, just give him a briefcase and tell him that his job depends on his winning an oil contract in a country that is not Norway. In this scenario, an executive from Apple Computer would not be dispatched to Boston to sell MacBooks to an eager client whose accounts are examined by the IRS and the SEC. Instead, he would be sent to Baku to win drilling rights to one of the fields that were up for grabs in the 1990s. The man from Apple would find himself in the crazy Skinner box that was Azerbaijan in the 1990s. It is impossible to understand oilmen if you do not understand the box.

In Azerbaijan, Western oilmen were competing for an initial round of contracts that involved paying at least $7 billion to develop offshore fields in the Caspian Sea from which a million barrels of oil would be extracted on a daily basis by 2010. Combined with contracts for other fields and a transnational pipeline, the total value of the contracts would reach tens of billions of dollars; the revenues from the oil would be substantially higher. The competition for the Caspian basin was billed as the last great oil rush of the modern era, though it would not be the first time a fortune had been made in Baku. In the late 1800s, France's Rothschild family and Sweden's Nobel brothers built their financial empires from the oil of Azerbaijan, which at the outset of the twentieth century produced half the world's petroleum. As the Soviet Union fell apart nearly a century later, the door reopened to foreign companies whose technology could extract crude that was beyond the reach of Soviet expertise. A crack team of behavioral psychologists could not have concocted a better environment for bringing out the worst in their human subjects.

The experiment in Baku, Azerbaijan's capital, was centered at the Intourist Hotel, which by the 1990s had not been burdened with much maintenance since its construction in the 1930s. The Intourist was the only place foreign oil executives were allowed to stay as they wined, dined and negotiated with ministers, middlemen and warlords who might possess the power to issue exploration contracts. The hotel, a five-story brute, was located on the aptly named Neftchiler Prospekt—Oilmen's Avenue. Tantalizingly, hotel residents could see, from their balconies, the derelict rigs that dotted the waterfront, visual cues to the huge fields that were farther offshore.

The oilmen did not have the Intourist to themselves. The hotel also played host to mercenaries from America and Afghanistan who were fighting in Azerbaijan's war with Armenia. In the early 1990s, this was the unfinished business of the Soviet collapse. And among the guests were diplomats, spies and journalists reconnoitering this newest and shakiest of nations. When the American, British, Turkish and Russian governments initially set up their embassies in Azerbaijan, all were located in the Intourist Hotel. (Because space was so tight, the British embassy began its operations in the offices of British Petroleum.) Everyone was trying to figure out what everyone else was doing, who was meeting whom, what was being said, who was telling the truth, who was working a con and who was spending time with the perfumed women of the evening in the basement tavern that some guests referred to as Ho Bar. There was an unseen witness to everything said—the Azeri security services, a thriving vestige of the KGB, which had wired the hotel from top to bottom. Going outside was one way to avoid listening devices, but at times it was too perilous to try that, because crime and political instability made the streets dangerous at night and, occasionally, during the day, too.

The representatives of BP, Amoco, Pennzoil, Unocal and lesser outfits squeezed into this squalid box. Because the hotel's services varied from sporadic to absent, the oilmen imported, on their corporate jets, things like breakfast cereals, printer paper and bottles of Johnnie Walker, so that they could enjoy a shot or four at the end of the day and not wonder whether it had been diluted with antifreeze by the waiters at Ho Bar. Phone lines were assumed to be tapped by the Azeri KGB or rival oilmen who wanted to know who their competitors were talking to and what they were saying. I visited Baku a few years ago and talked to oilmen who recalled the Intourist with an I-was-there horror—a visceral badge worn by survivors of momentous events. "We knew that all our rooms and all of our phones and all of our cars were bugged," one told me. "We would have a conversation in a car and within twenty-four hours a friend would come up and tell us exactly what we said. We would do silly things, too—we'd say we needed to meet one of the heads of a political party but we couldn't find him, and we'd just announce it in an empty room." Because the walls were bugged, the

announcement was heard by the hotel's eavesdroppers. "And twenty
minutes later, the manager of the hotel would say, Would you like to
meet with so-and-so?"

Sensitive documents could not be left in the rooms; everything of
competitive interest was carried in locked briefcases or protected by
guards standing outside the living quarters of executives when they had
meetings elsewhere. No one could be trusted, including the ministers
and bureaucrats who represented the government of the moment.
Until Heydar Aliyev consolidated his hold on power in the mid-1990s,
Azerbaijan endured a dismal reign of warfare, petty violence, utter
poverty and a succession of leaders who were no better than Mafia
chiefs. How do you negotiate in such circumstances? One of the representatives of the government was a shady Slovak businessman who was
never seen without a pistol strapped to his waist—and who once
pointed the gun at the head of an executive he was negotiating with.
Confidential bids leaked out almost the moment they were presented
to the government, because there was always a midlevel bureaucrat
who augmented his meager salary by showing company X's offer to
company Y. "You make a supposedly confidential proposal and the next
thing you know it has been shopped out by someone," an executive told
me. "It's several Rolaids a day, every day." Contracts would be signed
one day, bottles of champagne would be opened in celebration—and
within days the contracts would be declared null and a signing ceremony would be held with another company.

Like Alain, the executives I spoke with outlined the maddening
pressures weighing upon them yet denied any personal unethical activity. Their competitors, they added quickly, were breaking the rules all
the time. Apparently all oilmen in Baku were breaking the rules except
the ones who were telling me that everyone was breaking the rules.
What I knew for sure was that an Apple salesman would fail unless he
adapted his sales tactics to the horrible box that awaited him at the
doors of the Intourist Hotel.

The outside of the box looked different, of course.

On a February evening in 2003 I joined more than a thousand oil executives in a Houston ballroom that was large enough for a jumbo jet or two. The pin-striped diners were served plates of mixed salad, grilled salmon and chocolate mousse by overworked waiters whose service was as gentle as cowboys heaving bales of hay to livestock. This was the gala evening of an annual oil conference at the city's Westin Galleria Hotel, located in a massive mall that also featured a rink where Olympic skating champion Tara Lipinski trained. Drawn from across the globe, the men and just a few women in the chandeliered cavern constituted an oilpalooza.

The attraction on this evening was neither food nor skating but a chemical engineer from South Dakota. Since 1963 he had worked for just one company, eventually becoming its chairman and chief executive. He made everyone else in his hard-bitten industry seem gentle. He was gruff even to members of Congress and scoffed at global warming long after scientists proved it. Greenpeace called the company he led America's "number one environmental criminal." He was superficially unappealing too, with a misshapen lip, an ample belly and a set of jowls that cartoonists would judge absurd. But in the oil industry you do not need to be pretty or kind to succeed, and this oilman had succeeded beyond anyone's imagining. Lee Raymond had turned Exxon-Mobil into the largest and most profitable corporation in America. He was rewarded with an astounding $686 million in compensation during his thirteen-year tenure as chief executive, which breaks down to $145,000 a day, or more than $6,000 for every hour he worked, slept, ate or golfed.

Raymond fascinated me. Despite his stature and power, he was nearly unknown outside the environmental lobby, which despised him; the financial industry, which swooned over him; and the oil industry, which feared him. (Exxon's executive suite was known as "the God Pod.") Think of the tycoons who are part of the contemporary lexicon—Gates, Murdoch, Buffett, Jobs, Branson—and realize that absent from their ranks is the longtime leader of one of the most profitable multinationals of the twentieth century. Raymond was smart enough and secure enough to neither crave nor need publicity, which he knew would invite unfriendly questions. He did what he had to do, meeting financial journalists to announce earnings, but little more. He turned down my requests to interview him.

I wanted to see him because he was not just in the highest echelon of his industry's ruling class; he seemed its epitome. Oil firms employ millions of people as ditchdiggers, roughnecks, machinists, geologists, technicians, accountants, lawyers and consultants—a global army presided over by a commanding elite of larger-than-life executives. Like a nation or nationality, the industry has its particular belief system, its financial and political interests, its social layers and pecking orders. In some ways, it has the hallmarks of a political party *and* a religious movement. Use whatever metaphor you wish—senators, high priests or enforcers—it is impossible to know the oil industry without knowing the men who run it.

Mousse plates cleared, Raymond lumbered onto the ballroom stage. The crowd offered a round of applause that was akin to a handshake rather than a hug. In this industry, there was no need to feign love; grudging respect would do. Raymond's lumpy, uneven physique imparted an off-the-rack look to his tailored suit. He made not a single attempt at humor, and he uttered every word with a metronomic drawl. He felt no compulsion to entertain or please.

His speech was an industrial mission statement. His listeners, who included government ministers, princes and CEOs, were reminded of how vital their work was, how underappreciated they were, how they must labor harder than ever, how the future will be grander than the already blessed present. A video screen enlarged Raymond's presence to superhuman proportions. It was part Tony Robbins, part Billy Graham, with a whiff of a mumbling Leonid Brezhnev. Invoking a sacred industrial purpose, Raymond recited his version of the inspirational commandments of the oil world:

"We all have a tremendous opportunity and a responsibility to improve the quality of life the world over. Virtually nothing is made without our energy and our products."

"Our industry's best years lie ahead, surpassing even the greatest achievements of the century gone by."

Lee Raymond earned $686 million in his thirteen-year tenure as chief executive of ExxonMobil.

"We condemn the violation of human rights in any form, and believe our stand on human rights sets a positive example for countries where we operate."

"It is almost impossible for someone who is not in the industry to begin to understand the magnitude of the industry and what we do."

The audience's reaction was ritualized, less a genuine wave of applause than an obligatory simulation. I was reminded that in this brutal industry, it was best to save your enthusiasm for crushing a rival rather than congratulating him.

Greed

The court building in Odessa, Texas, is a nearly windowless horror, designed in the 1960s by local architects who followed the Brutalist style popular at the time. If the building could talk, it would say there is no such thing as too much concrete or too little sunlight. I stepped inside on a summer day, seeking illumination that was neither solar nor fluorescent. I wanted to learn about the oil industry's tendency to cut corners even in its own backyard.

Odessa is at the center of the Permian Basin, the great oil reservoir of west Texas. The city, with a population of nearly 100,000, has a love-hate relationship with its roughneck past. The cliché is that if you want to raise kids, do so in nearby Midland, and if you want to raise hell, do it in Odessa. When I visited, the town was less wild than it used to be, focusing more on economic development and high school football than tequila nights that ended in brawls and broken teeth. A museum, dedicated to all things presidential, emphasized the local oilmen who became the forty-first and forty-third presidents, George H. W. Bush and George W. Bush; visitors could pick up complimentary maps pointing them to the Odessa and Midland homes where the Bushes lived. Midland had proudly created its own Petroleum Museum, which claimed in its brochures to be "the nation's largest museum dedicated to the petroelum industry and its pioneers." If there was a place Big Oil might call home, where it would behave with conscience, it would be Texas, its petromanger.

Workers on an offshore oil platform

At its peak in the 1960s and early 1970s, output in the Permian Basin was 1.8 million barrels a day, pulled from a swath of mesquite plain that measured roughly 250 by 300 miles. Back in those days, west Texas was the Saudi Arabia of oil. At one point, the bonanza reportedly turned Midland's Rolls-Royce dealership into one of the world's busiest. But as the Permian Basin passed its prime, the big companies moved on, selling out to smaller firms. By the 1990s, with prices and output tumbling, the office towers in Midland that had once housed Exxon, Mobil, Gulf and Shell were vacant, not a tenant anywhere—tombs touching the sky. Those hard years turned bitter when people concluded that "the majors," as the big companies were known, had stolen from them.

It was at the county clerk's office in the Odessa court building that I learned about the alleged plundering. A lawsuit filed in 2003 accused Exxon, Chevron, Shell and nearly a dozen other firms of a decades-long conspiracy to avoid royalty payments on oil they extracted from public land. In the nineteen-page complaint, the words "fraud," "fraudulent" and "defraud" appear more than two dozen times. I was accustomed to hearing of predatory corporate behavior in Africa and the Middle East, but this was startling—a major lawsuit against Big Oil in its backyard, and the suit wasn't the work of Greenpeace. It had been filed by the county government, with similar suits filed by nearly a dozen other municipalities in the Permian Basin. *Texans* were accusing Big Oil of stealing from *Texas*. And the accusers weren't Austin liberals but oil-patch Republicans, as pro-business as possible.

The alleged fraud was simple. Companies pay about 12 percent in royalties on oil extracted from public land. The royalties are pegged to the price of oil when it is extracted—the "posted price." The suit accused the firms of selling oil for more than the posted price, thereby avoiding royalties on the price differential. "Defendants knew that their material misrepresentations regarding the market value of oil were false at the time they were made," the suit said. "The representations were willful and malicious." This unfriendly language was chosen not by Ralph Nader but by lawyers like Russell Malm, the county attorney for Midland. As a Republican, Malm's first impulse had not been to sue an oil company, but after looking at the data he'd realized, as he told me when we were chatting in his office, "The oil companies didn't come out here just to do good for the Permian Basin."

It can escape notice that the United States, the world's largest importer of oil, is also the third-largest producer—some 8 million barrels a day. This provides a window for evaluating the core principles of the firms that dominated the global industry for most of the twentieth century. Overseas, American firms needn't try hard to get the better of inefficient and corrupt governments, and they don't suffer the restraint of cultural bonds to the societies where they operate. The practical and ethical barriers to cutting corners are lower overseas. Yet even in America, as the Odessa case and a multitude like it showed, oil companies have a remarkable tendency to do their best to get around the law.

Businesses cheating in America—this is not news. There is hardly an industry in America that hasn't been indicted for something. Even the paragon of cool, Apple Computer, was investigated for backdating stock options for its hip CEO, Steve Jobs. But there is a startling pattern in the oil industry; in Texas, Alaska, Louisiana, California and elsewhere there has been an unending stream of indictments, trials and fines for oil companies breaking the law on royalty payments, environmental protection and worker safety. This carnival of sin dates back at least a century, to when the Supreme Court ordered the breakup of John D. Rockefeller's Standard Oil, which was ruled to be a price-fixing monopoly that used blackmail and bribery to get its way. Industry practices were not greatly improved by that verdict. American oil firms and executives traded with and supported Nazi-era Germany even after World War II began. As Interior Secretary Harold Ickes wrote in his diary at the time, "An honest and scrupulous man in the oil business is so rare as to rank as a museum piece."

Cheating continues in America—lawsuits, fines and settlements have not abated in recent years—but the greatest swindles tend to occur overseas, where the dynamic is reminiscent of doping scandals. Sports authorities develop laboratory tests to detect performance-enhancing drugs, and athletes respond by using new drugs that are not detectable by the latest tests. In the oil and gas world, as in track and field, the one-step-ahead-of-the-law violations are intentional rather than accidental, as the case of Jeffrey Tesler showed.

In 1977, Congress passed the Foreign Corrupt Practices Act, which criminalized bribery of foreign officials. For nearly two decades, the law netted mysteriously few violators; annual prosecutions could be counted on one hand. One of the reasons, aside from indifference on the part of federal prosecutors, was that oil companies had begun to outsource bribery to middlemen or joint-venture partners. On occasion these missing links were discovered, and this happened in a spectacular fashion in 2003, when French regulators looked into an irregular series of payments totalling $132 million by a consortium of international firms that was bidding on a multibillion-dollar natural gas project in Nigeria. The consortium was led by M. W. Kellogg, which was then owned by Halliburton.

The payments reportedly went to Jeffrey Tesler, an obscure lawyer who worked in an immigrant neighborhood of London and whose clients included sex shops with zoning problems. Why would a global energy consortium funnel money to a barrister whose office, adjacent to a Somali butcher, advertised the availability of a fax machine for fifteen pence a page? In the 1980s Tesler had helped members of Nigeria's nefarious elite buy homes in Britain. This meant Tesler was connected to Nigerians who could probably influence government contracting decisions. As it turned out, the consortium, which included firms from Italy, France and Japan, had agreed to pay Tesler $60 million if the consortium won the contract. Tesler and a consortium official allegedly agreed to channel $40 million of this fee to General Sani Abacha, the Nigerian military ruler who, after his death, was reportedly found to have stolen more than $3 billion from government revenues. The consortium won the contract and Tesler allegedly received the $60 million payment and substantial additional ones, until the total was $132 million.

This scheme was not discovered by a compliance officer working for one of the consortium members. Major corporations employ lawyers to ensure that contracts and payments comply with laws like the Foreign Corrupt Practices Act; they are known as compliance officers. Unfortunately, compliance officers in all four companies failed to notice the payments or, noticing it, did not object. French prosecutors heard of the scheme only after filing embezzlement charges in an unrelated case against an employee of a French company in Halliburton's consortium. The employee, upset that his firm refused to defend him in the embezzlement case, took his revenge by telling French authorities about the payments to Tesler—and that is how the scandal began to unravel. In 2009, Halliburton admitted its role in the affair and agreed to pay more than $550 million in fines to the U.S. government. Albert "Jack" Stanley, the Halliburton executive who oversaw the corruption scheme, pleaded guilty to violating the FCPA. Tesler has been indicted and faces trial in the United States. It was one of the largest bribery scandals in American corporate history, and it was uncovered only by chance.

As I studied the industry's global rap sheet, I wondered about the reasons for this pattern of law evasion. Why did oil firms transgress legal and moral strictures on such a consistent basis?

Oil firms describe their work as "producing" oil, but in truth they "produce" nothing, insofar as the meaning of the word is generally understood. With permission from host governments, they extract a valuable liquid from the earth. The process of drawing it from the ground, removing impurities and shipping it to refineries is complicated and daunting, but profits depend first and foremost on getting permission to extract the valuable stuff. It is not the customer who is king and determines a company's destiny but the president, minister, senator, mayor or real-life royalty who grants extraction licenses. If the computer industry operated this way, Dell and Samsung would bid for the right to pull semiassembled laptops from under the deserts of Arabia or the swamps of Africa. Especially in times of high prices, the dynamic of extracting rather than manufacturing helps explain why oil firms have a record of bribing foreign officials. It is the same in other extractive industries, such as mining for gold. The permission from the host government is what matters, because the product will sell itself, usually with wide profit margins.

If the world's resources were all located in Norway and Canada, where the national governments are strongly resistant to bribery, corporate corruption would not be a severe problem because there would be few takers for under-the-table offerings. Firms would have to win contracts in open competitions and pay fair royalty rates and operate their concessions in a responsible way, because well-funded regulators would strictly enforce laws that had not been watered down by lobbyists. But Norway and Canada are outliers; most of the world's oil and gas resides in countries with bribery-prone systems, and even the United States fits into that category, as illustrated by a corruption scandal in 2008 that involved a number of royalty officials in the Interior Department. In countries like Nigeria and Iraq, closed-door negotiations for extraction licenses and legislative votes on royalty rates turn into matchmaking opportunities for bribe givers and bribe takers. Nordic ministers might not think of accepting money, but a Caspian dictator would be inclined to demand and take it in an instant.

Unlike Intel, which is not masochistic enough to build a factory in Equatorial Guinea or Kazakhstan, Exxon must do business in such places. Geology determined where oil is located and where, therefore, Exxon must operate. When other industries operate in morally dubious countries, corners tend to get cut, too. Microsoft bent to the demands of China's government by banning the use of words like "democracy" and "human rights" on its search engine and blogging service. Yahoo knelt further, helping Chinese authorities identify a democracy-promoting reporter who used its e-mail service; the reporter was jailed for ten years. The difference is that extractive industries do most of their business in compromise-inducing countries, in a sector with structural incentives for corruption, and they have vast footprints that alter political, economic and environmental destinies. Microsoft's compromises might be distasteful, but they do not contribute overtly to violence and poverty.

The problem is not that extractive industries have lower principles than other industries. The problem is that they must have better principles. Unfortunately, having a soul is a luxury the law and shareholders do not encourage.

If Lee Raymond, the legendarily coldhearted chief executive of Exxon-Mobil, had arrived at work one morning in an altruistic stupor and ordered that half of the company's profits be devoted to social-welfare programs in the Niger Delta or decided that the company's lobbyists should support higher royalty rates in Texas, his board of directors would have responded by issuing a statement explaining that Raymond was taking a leave of absence to spend more time with his family. A truthful statement from the board would admit that the lunatic had to be terminated. The board would have had little choice, thanks to rulings that began with a 1919 judgment on a dispute between Henry Ford and the Dodge brothers, Horace and John.

Ford, who owned the majority of shares in Ford Motor Company, decided to suspend special dividend payments so that more funds would be available for capital investment as well as price reductions. He also wanted to prevent the Dodge brothers, who were minority shareholders, from amassing enough capital to move forward with their plan to set up a rival auto firm. The Dodges filed a lawsuit demanding the dividends. In testimony, Ford made a surprising argument—that his company's goal was "to do as much good as we can, everywhere, for everybody concerned . . . and incidentally to make money." The Michigan Supreme Court would have none of it, ruling that a corporation's mission "is organized and carried on primarily for the profit of its shareholders." Ford was ordered to pay.

The *Dodge v. Ford Motor Co.* ethos goes back a ways. In the 1700s, Lord Edward Thurlow famously lamented that under the laws of his day, "Corporations have neither bodies to be punished nor souls to be condemned." A modern-day justification for conscience-free companies is nearly paradoxical: that they do good by not trying to do good. Milton Friedman championed this notion in a famous article entitled "The Social Responsibility of Business Is to Increase Its Profits," which argued that companies interested only in profits do the greatest amount of good by creating jobs and returning dividends to their mom-and-pop shareholders. Friedman and other conservatives did not oppose all forms of corporate generosity, and nor does the law, because companies can enhance their image and thus their sales by funding summer camps and literacy programs. But generosity that offers no payback—this is discouraged, legally and ideologically. Altruism is fine so long as it isn't true altruism.

I witnessed the consequences when I met a Chevron executive at the company's headquarters in Lagos. I had to make my way through dozens of protesters who had traveled several hundred miles from their fishing village to throw onto the floors of Chevron's reception area some of the oil that had contaminated their waters. I was escorted around the protest to the executive's office, where I asked whether Chevron might do more to help the villagers and promote general welfare. I noted that a few days earlier the firm had announced quarterly earnings of $3.2 billion. "It is not the role of an oil company to provide the basic fundamentals that are required for communities to thrive," the executive replied. "It is not our mission as a corporation. It is not our identity as a corporation." Regarding the protesters who were blocking the reception area and singing a song with the refrain "We have suffered enough in the hands of Chevron / And we cannot continue to suffer like this," he said the company would look into their grievance.

One day, I got a behind-the-scenes look at a board of directors meeting that was a sort of corporate vaudeville.

I was in Moscow working on a profile of Vagit Alekperov, the president and a multibillionaire shareholder of Lukoil, a Russian company that has oil reserves equal to Exxon's. The boardroom at Lukoil's headquarters had wood-paneled walls and parquet floors that had been polished to an opulent shine. The board members, all but one of whom were men, sat around an oblong conference table with enough space for several dozen people. It was dotted with bottles of Evian and porcelain coffee cups replenished by waiters who slipped into the room like silent ships. The two foreigners on the board, Richard Matzke, a former Chevron executive, and Mark Mobius, a financier, were emblems of Alekperov's effort to globalize Lukoil and list its shares in London and New York. I had been invited to watch a model board meeting.

As the meeting began, there was a malfunction with the headsets providing an English translation of the proceedings. When the problem was fixed after a few minutes, the Lukoil executive who was running the meeting said, "I think our [American] colleagues missed very little." That was true. And they missed very little in the PowerPoint slides projected onto a screen without translation of the Russian captions. Vague spending plans were outlined by several executives, a few investments were described and one or two expansion opportunities were mentioned. It was as numbing and unrevealing as an annual report without the glossy photos.

Alekperov, at the front of the room, spent much of his time reading a company brochure. When he wasn't doing that, he fiddled with a fountain pen; he spoke only once. After a perfunctory vote to approve everything that had been outlined by the management, the meeting was wrapped up with final words from Valery Grayfer, chairman of the board but far lesser in wealth and status than Alekperov. "I thank you all, dear colleagues," the gray-haired Grayfer said meekly. "Our work is finished today." The meeting had lasted for less than an hour.

Shows of this sort are common at board meetings. Particularly in the oil world, CEOs have for the past century been especially strong-willed, accustomed to running things with minimal oversight or inter-ference from their boards (the members of which the CEOs usually handpick). So long as company executives are sustaining or expanding profits, boards will give a nod to almost anything that comes up in their meetings, including exorbitant compensation packages. (Exxon's board did not blink as Lee Raymond walked away with that $686 million in his CEO years.) Spirited arguments about oil pollution, human rights violations or corruption in the countries where contracts are signed will perhaps be made by dissident shareholders who grab a microphone at an annual shareholders' meeting, but that's usually it. Board meet-ings are intended to ratify the status quo, not disturb it.

A look at board membership helps reveal why. In 1991 Condoleezza Rice joined the Chevron board. She had worked as a Russian expert on the National Security Council under President George H. W. Bush and at the time of her Chevron appointment was a professor at Stanford University. Just thirty-six years old, she had almost no experience in the financial world. But she had political connections. Shortly after joining the board, Rice traveled to Kazakhstan to help Chevron win a slice of the multibillion-dollar contracts being negotiated there. Chevron, in addition to paying Rice $60,000 a year for her part-time efforts and awarding her several hundred thousand dollars in stock, even named a supertanker after her (though the S.S. *Condoleezza Rice* was quietly renamed when its namesake became President George W. Bush's national security adviser).

Today, Chevron's board, like the boards of its rivals, consists of a friendly group of establishmentarians from the business, political, mil-itary and academic sectors. And they are drawn from both sides of the mainstream political spectrum. Until President-elect Barack Obama named him as national security adviser, retired General James L. Jones was on Chevron's board. Like Rice, Jones severed his corporate ties to serve in the White House. That doesn't imply a severing of sympathies or interests. Supportive of business as usual when they were on the board, neither Rice nor Jones showed a significant change of heart when they returned to government service. For a corporation, the only thing better than having a former White House official on its board is having a future official on it.

The board meeting I attended was a private vaudeville. The public vaudeville occurs at times of high gasoline prices, when Americans join Nigerians and Angolans in beseeching oil companies to sacrifice profits for the public good.

The script sounds serious. Inundated with pleas for relief at the gas pump, the companies respond that they are investing in exploration and doing everything they can to bring cheaper gasoline to the marketplace. A refrain is heard from the boardrooms in Houston and Dallas: *We are your friend.* As prices soared to ever-higher levels in recent years, industry advertisements in the *New York Times* and other newspapers offered reassurances that, as an Exxon ad put it, "Today's energy industry earnings are important for meeting tomorrow's energy needs." In 2006 Exxon spent $19.9 billion exploring for oil and updating its refining systems—an impressive number that was relentlessly promoted. The same year, without any mention in its self-lauding publicity, Exxon dispensed more than $35 billion on dividends and stock buybacks. The bulk of its windfall went into the hands of the company's owners.

This is completely normal in the business world. Apple, enjoying record earnings from its iPods, did not respond by giving away the wonderful little machines. It continued to sell them for as much as the hungry market would bear. Nobody expected that Steve Jobs would forgo his profits or invest them in socially useful projects. But when oil firms began to clock quarterly profits in the billions and then tens of billions, the public and politicians clamored for these companies to behave like nonprofits. The oil companies, preferring PR to honesty, did their best to portray themselves as public-minded.

The truth lay elsewhere. The chartered purpose of American oil companies is not to supply consumers with cheap gas but to make as much money as they can by selling their product at the highest possible price. It is reasonable to ask them to obey the law, but it's a different thing to ask them to have a soul and care about our pain. On occasion a CEO might admit the truth, but that's only because he wandered from the script. This happened on a morning in 2006 when Rex Tillerson, who succeeded Lee Raymond as Exxon's chief executive, was interviewed on the *Today* show. Softball questions were the norm on this show, and Tillerson certainly intended to polish Exxon's image, but the host, Matt Lauer, threw a curveball.

"Would ExxonMobil be willing to lower profits over the summer to help out in this time of need and crisis?" Lauer asked.

"We work for the shareholder," Tillerson said. "And the investors who own our stock are over two million Americans. A lot of pension plans, a lot of teacher retirement plans, and our job is to go out and make the most money for those people so their pensions are secure, so that they see the benefits of our work."

Lauer wasn't satisfied.

"That's a no?" he asked.

"We're in the business to make money," Tillerson replied.

If the industry's critics were hoping for a paradigm-changing chief executive, John Browne seemed to answer their dreams. A gregarious physicist who loved cigars, art and opera, Browne became chief executive of British Petroleum in 1995 and started a revolution. Everything Big Oil had done wrong for the past century, BP would do right, he vowed. As Exxon continued to fund pseudoscientific groups that claimed global warming was a hoax, Browne promised to cut BP's carbon emissions and spoke in favor of the Kyoto Protocol, which most oil companies vehemently opposed due to the expected onset of carbon caps and taxes. The American Petroleum Institute, a lobbying arm of the industry, sourly complained to Browne that he had "left the church."

He was making a new one, launching a $200 million rebranding campaign in which British Petroleum's name was shortened to "BP" and its logo became a green-and-yellow sun—a happy friend of the earth. The firm ran TV commercials that extolled its solar and biofuels programs and often used the slogan "Beyond petroleum." Of course, BP continued to depend on fossil-fuel revenues, so its ads acknowledged the step-by-step nature of corporate change with the catchphrase "It's a start." In essence, BP asked the public to trust its sincerity when it promised to be as green and conscientious and forward-looking as possible.

Skeptics were naturally concerned that BP was engaged not in revolution but in greenwashing—using climate-friendly PR to make the public forget climate-unfriendly realities. For instance, while the company announced that it was best friends with Mother Nature, it supported efforts to allow drilling in Alaska's Arctic National Wildlife Refuge. Greenpeace mockingly gave Browne an award for "Best Impression of an Environmentalist," but many environmentalists quietly hoped that he meant what he was saying.

It didn't take long for them to get an answer.

In 2005, a BP refinery in Texas suffered a massive explosion that killed fifteen workers and injured hundreds. Investigations revealed that BP had cut the refinery's capital budget by 25 percent. Broken or outdated equipment had not been replaced, while worker training and safety had been ignored. Months before the explosion, the refinery had commissioned an independent report that had warned, prophetically, of "a series of catastrophic events." The report's authors wrote, "We have never seen a site where the notion 'I could die today' was so real." Even though BP earned a record $22.3 billion profit in 2004, the refinery ran on a catastrophe-beckoning budget. A BP official admitted that the disaster had been caused by "incompetence, high tolerance of noncompliance, inadequate maintenance and investments." BP set aside $2.1 billion to settle lawsuits related to the explosion. In the twenty-first century, a Dickensian tale of greed and callousness had unfolded at a facility owned by a firm proclaiming itself the most conscientious its industry had known.

This was not the end. A year later, a BP pipeline dumped more than 200,000 gallons onto the North Slope region of Alaska's coast—the largest spill ever on the slope. BP, ordered by regulators to inspect the entire pipeline, found that corrosion had reduced the pipeline's steel in some places to the thickness of a beer can's shell. The pipeline was closed for urgent repairs. Follow-up investigations found that the company had cut costs by forgoing maintenance and updates. As one newspaper wryly noted, "For a company that claims to have moved 'beyond petroleum,' BP has managed to spill an awful lot of it onto the tundra in Alaska."

BP was turning into an example of iniquity rather than virtue. A few months after the spill, BP traders were indicted for fixing propane prices. The indictment quoted a BP trader as saying, in a recorded call, "What we stand to gain is not just that we'd make money out of it, but we would know from thereafter that we can control the market at will." The indictment said that a dry run had even been conducted by the traders and that the scheme, which raised home heating prices in the winter, had "the knowledge, advice and consent of senior management."

It became possible to suggest that BP stood for Beyond Parody, except that the firm's self-impalement was not yet finished.

Lord Browne's sexual orientation—gay—though not a secret, had not been discussed publicly. In 2002 he began a relationship with Jeff Chevalier, a Canadian student. When the relationship ended, Chevalier asked for $600,000 to soothe the hurt. Browne refused, Chevalier went to a tabloid and Browne sought an injunction. Browne lost the case and, problematically, lied in his testimony, claiming that he'd met Chevalier while exercising in a park, rather than through suitedandbooted.com, a gay escort Web site on which Chevalier's services were advertised. As the presses rolled, Browne resigned. Booted, indeed.

Browne was a torn oilman. He probably would have liked to make BP clean, green and safe, but he needed to satisfy the market, too. "Corporations have to be responsive to price signals," he once said. "We are not public service." This was the sort of remark Wall Street likes to hear. Yet his cost cutting was a key factor in the explosion and spill. In the end, Browne could not make BP a friend of the earth and a friend of the market.

Big Oil is getting the reward it deserves: after more than a century of power and indecency, it is shrinking.

Until the 1970s, Western companies controlled most of the world's oil and gas. Today, thanks to nationalism and nationalization, Western firms control less than 15 percent of world reserves, and their grip erodes further every day. The bulk is now in the hands of state-controlled companies like Saudi Aramco, Gazprom, Petróleos de Venezuela, National Iranian Oil Company and China National Petroleum Corporation. Exxon, the largest oil company in America, does not even rank in the top ten globally in terms of the oil and gas reserves it controls. State-owned firms, known as "national oil companies," now set the rules; once-mighty Western companies are being turned into contractors rather than owners. Unfortunately, this is not necessarily an improvement.

With noble exceptions in Norway, Saudi Arabia and Brazil, state-controlled energy companies, though not obliged to follow the fiduciary logic of *Dodge v. Ford Motor Co.*, tend to do as much or more harm than their profit-seeking counterparts on the New York Stock Exchange. The dirtiest oil facilities I have seen were run by state-owned Petroecuador, a toxic example of how a poor government cuts corners on environmental and worker safety. Even Petróleos de Venezuela, run by Hugo Chávez's leftist regime, has an unenviable environmental record, because much of its revenues go into the government's social programs, some of which are useful, others of which are wastes of money. Angola's state-controlled oil company, Sonangol, has been used as a piggy bank by corrupt officials.

The ethical practices of Exxon and Chevron might even look good when compared with those of Russia's Gazprom or China's CNPC, which have become major global players in just the last decade. Although it's hard to imagine the energy business becoming more competitive, politicized or secretive, that's happening as global power is defined by control of energy reserves. The U.S. government prohibits American firms from operating in Sudan, whose regime is responsible for genocide in Darfur, but China is delighted to send its companies there. Sudan's oil is now extracted mainly by Chinese firms that are not restricted by laws like the Foreign Corrupt Practices Act. Although Western firms were hardly transparent about their operations, state-controlled companies tend to be far more opaque. If the profit motive led to unethical behavior by Western firms, imperatives of national glory (or national survival) can be worse incentives for state-controlled ones.

It is comforting to think of history as progress, of life getting better, maybe not every year but over the course of time. If that were so, the future would be dominated by oil and gas companies that shun bribery, that genuinely care about the communities and countries they operate in, that refuse to deal with dictators and thieves. Yet the threat is that the future will belong to state-controlled corporations whose behavior may make us nostalgic for the trifling days of cheating on royalties in West Texas.

In Baghdad, the Ministry of Oil turned into the Ministry of Truth, or so it seemed in the spring of 2003. While most government buildings, including the National Museum, were looted of everything from artwork to computers and light bulbs, after which the remains were often set alight, the Oil Ministry, as I said earlier, was untouched, aside from a bit of vandalism in the hours between the melting away of Saddam Hussein's regime and the arrival of American soldiers. This was not a matter of luck.

When I visited the cream-colored ministry one morning, a concrete plaza at its entrance was sealed off by coils of barbed wire and scores of marines equipped not only with .50-caliber machine guns but a particular view of current events. There were no roving bands of looters near the ministry, nothing more threatening than echoes of chaos rather than chaos itself, so Corporal Shane Evans had enough time to show off a tattoo that read, "Anyone who sees through death becomes ageless, deathless and immortal." He also displayed a crafted view of the relationship between oil and invasion, when I asked about it. "There's a misconception," he said. "We're just coming here to give Iraqis freedom."

His duties seemed at war with his words. He was not defending one of the hospitals or schools being stripped bare, nor stopping the carjackings and shootings, nor chasing down museum thieves or the fugitive Saddam. These were not his orders. The nerve center of Iraq's oil industry never ceased to receive protection from Americans like Corporal Evans, and this made Iraqis suspicious.

"The Americans will not steal the oil but they will control it; they will pull the strings," said a ministry official for whom the future remained uncertain enough that he offered only his first name, Mohammed. He was standing with dozens of colleagues who were seeing one another for the first time since shock and awe had morphed into looting and burning. They hugged like the beached but relieved survivors of a sunken vessel of state, and one of them touched my arm to remind me, "It is all about oil."

When oil comes from the ground, as I've said, it is filled with impurities that include dirt, water, salt, arsenic and mercury. This is crude in its raw, literal form. The refining process transforms this black swill into a clear fluid without which our civilization would collapse. Quite often a corollary process of political refining occurs to sanitize the truth of what's done to keep oil in the hands of friendly governments. Just as cars cannot run on unrefined crude, political systems choke at the unfiltered mention of war for oil. It is difficult for states-

The Ministry of Oil in Baghdad on the day American troops took control of Iraq's capital

men to talk honestly about what they do or don't do for oil, and so the facts are shaded by euphemisms and lies. As a result, everything and nothing are attributed to our desire for this filthy, vital substance.

President George W. Bush insisted before the invasion that it had nothing to do with oil, that it was about weapons of mass destruction and, to a lesser extent, democracy. He was not being honest. I followed American troops into Iraq to learn what I could from actions rather than speeches. It was easy, standing in front of the Oil Ministry, to accept what seemed so obvious, but after several months in Iraq I realized how confounding oil can be.

The marines who surrounded the Oil Ministry waved me inside. Ground-floor windows were sandbagged, bathrooms emitted a stench of clogged dysfunction, floors were littered with papers and fallen tiles, and a flashlight was needed for navigating internal corridors that were darkened in the post-Saddam era of now-you-see-it-now-you-don't electricity. The ministry was part fortress, part sarcophagus. Computers that hadn't been stolen had been switched off to avoid ruination by power surge, and the phones didn't work because the city's telecom network was an early casualty of American bombs. A few days after the old regime had fallen, a new one had not yet taken its place, even in this center-of-all-things locale. While I waited for the future to arrive, I had time to look the past squarely in the eye—literally.

On the ground floor, just past the glass doors that let you into the ministry, there was a small lobby that spilled into two wide corridors, one on the right, the other on the left. If you walked straight ahead, toward the back of the lobby, you would approach a metal bust of Saddam. It had rested for years in the middle of the lobby on a chest-high pedestal, from which the dictator glared by bronze proxy at all who entered the ministry. The bust, like Saddam, was no longer on a pedestal. It had been taken down and dragged across the marble floor. Its nose nearly touched the wall, so that Saddam, now at knee level, was staring at no one. A soiled carpet had been thrown over his head, bestowing a final insult. Like Iraq's petroleum, the statue had no voice, though it had a story to tell.

Saddam's rise to power began in a 1968 coup led by General Ahmed Hassan al-Bakr, who relied on the help of his young and ambitious cousin from Tikrit. In 1979, after shoring up behind-the-scenes powers, Saddam forced al-Bakr to resign and became president of Iraq as well as secretary-general of the Ba'ath Party. In a famous meeting a week later, Saddam read the names of more than sixty-eight party members who, he said, were spies or traitors. One by one they were led out of the room; twenty-one of them were executed. The event was filmed, so that all of Iraq would know to fear their new leader, who was a natural-born thug from the first day. The Internal Security Services, the Mukhabarat, would execute thousands of Iraqis in the years to come and terrorize the nation as well as its neighbors.

In 1980, Saddam ordered his army to invade Iran, and during the eight-year conflict that ensued, one of the bloodiest since World War II, his troops repeatedly used chemical weapons. This did not bother the great powers. The Reagan administration was officially neutral in the war but tilted toward Iraq's side in the tradition of the enemy-of-my-enemy-is-my-friend philosophy; Ayatollah Ruholla Khomeini's Iran was regarded as a greater threat. The United States even supplied Iraq, via intermediaries, with intelligence about Iranian troop locations. Washington did not supply arms but made sure Saddam had what he needed, including cluster bombs the CIA arranged to be sold to Iraq by a Chilean firm. President Reagan even dispatched a special envoy to tell Saddam that America wanted to improve relations. Donald Rumsfeld, a pharmaceutical executive at the time, shook Saddam's hand and chatted with him for ninety minutes. When Saddam later fired mustard gas and nerve agents at his own people, the Kurds in Halabja, the Reagan administration barely protested and tried to shift blame away from Saddam. "There were indications that Iran may also have used chemical artillery shells in this fighting," a State Department spokesman said a week after the attack.

Saddam was a menace of the first order, but he picked his fights with America's principal enemy and sold his oil to Washington, so there was no need to punish him and there were several reasons to reward him. That is why nearly everyone was surprised when Iraqi tanks moved into Kuwait on August 2, 1990; in the classic political configuration, Saddam was our guy and was not supposed to obliterate one of our other guys. The U.S. ambassador to Iraq, who had met with Saddam a few days before the invasion, was on vacation, and President George H. W. Bush got the news while receiving heat treatment for his sore shoulders after golfing in Kennebunkport. But everyone realized immediately what was at stake.

Kuwait possesses 101 billion barrels of oil, about 8 percent of the world's reserves. With Kuwait's reserves added to Iraq's, Saddam would control one-quarter of the world's oil and become the most powerful Arab leader of modern times. More tempting, his army could move from Kuwait into eastern Saudi Arabia and seize its fields. Even the implicit threat of invasion might compel the Saudi royal family to heed Iraqi wishes. Saddam did not need actual control of Saudi fields to become the master of nearly one-half of the world's oil. His invasion of Kuwait proved the substance's allure: at the risk of ruin, even countries with an abundance of oil desire more.

In his memoir, the first President Bush recalled that his initial reaction was for "our vulnerable friend Saudi Arabia." British prime minister Margaret Thatcher had similar fears. "They won't stop here," she advised Bush. "They see a chance to take a major share of oil. It's got to be stopped." Dick Cheney, the secretary of defense at the time, had oil on his mind, too. "[Saddam] has clearly done what he has to do to dominate OPEC, the Gulf and the Arab world," he told Bush. "He is forty kilometers from Saudi Arabia and its oil production is only a couple of hundred kilometers away. If he doesn't take it physically, with his new wealth he will still have an impact and will be able to acquire new weapons."

On both sides of the front line, this was a war for oil.

There are lots of ways to fill up your car, a multiplicity of service stations to choose from. Similarly, there are lots of ways the American government gets the oil its people require. War is a last resort, when other tactics have failed; in a way, war is gasoline by other means.

The American government does not buy or transport the foreign oil we consume. That work is left to oil companies. The government has a set of far tougher tasks. First, it tries to ensure that oil reserves are controlled by friendly governments and friendly companies that will extract and transport steady supplies of oil to us. The government prefers that American companies be the ones—this creates profits and jobs for America itself and lowers the pulse rate of national security officials who worry about cutoffs by foreign companies, including European ones, that might be persuaded, for economic or political reasons, to ship their product to other countries or not ship it at all. As a result, the government lobbies intensively for extraction contracts to be awarded to American firms (as the British government does for BP, and the French for Total).

Second, the government tries to ensure that the infrastructure exists for getting the stuff from there to here. In particular, the construction of multibillion-dollar transnational pipelines that determine whether, for example, landlocked Kazakhstan has the ability to send its billions of barrels of oil to America or to China. And of course the U.S. Navy patrols the Persian Gulf and other sea-lanes to ensure the safe passage of supertankers that deliver much of the oil we consume. That's why the U.S. military has been called an oil-protection service.

The only oil the U.S. government buys and holds, apart from the prodigious amounts used by the armed forces every day (if the Department of Defense were a country, its daily consumption of oil would be roughly equal to Sweden's), is for the Strategic Petroleum Reserve. To soften the blow of interruptions due to war, boycotts or natural disasters, the government decided to set up the SPR in 1975. The SPR, which consists of four giant salt caverns turned into reservoirs near the Gulf of Mexico, contains more than 700 million barrels of oil—enough for about thirty-four days of U.S. consumption.

Ideally, the government's tasks are accomplished behind the scenes and without warfare. But if persuasion and intimidation fail, the brass knuckles must be used. This has been true for some time, and it was confirmed explicitly by President Jimmy Carter in 1980, in response to concerns about Soviet inroads in the Persian Gulf. "Let our position be absolutely clear," he said in his State of the Union address. "An attempt by any outside force to gain control of the Persian Gulf region will be regarded as an assault on the vital interests of the United States of America, and such an assault will be repelled by any means necessary, including military force." The Gulf War of 1990–91 would not be the first time America employed violence to secure its oxygen of oil from the Middle East.

Mohammed Mossadegh, the prime minister of Iran a half century ago, was doomed.

His rise and fall began in 1951, when Iran's parliament nationalized the Anglo-Iranian Oil Company, a precursor to British Petroleum that was owned by the British government and paid more in taxes to London than to Tehran, even though the entirety of its oil was extracted from Iranian fields. In the annals of developed nations stealing from undeveloped nations, Britain's conduct in Iran during the first half of the twentieth century might be second only to the infamous rape of the Congo by King Leopold's Belgium. As the writer Stephen Kinzer noted, "The wealth that flowed from beneath Iran's soil played a decisive role in maintaining Britain at the pinnacle of world power while most Iranians lived in poverty." Even British foreign secretary Ernest Bevin admitted that without Iranian oil there would be "no hope of our being able to achieve the standard of living at which we are aiming in Great Britain."

Mossadegh, who became prime minister after the nationalization vote, quickly enforced it. Iranian troops and officials seized Anglo-American facilities, including the massive refinery in Abadan. The British government, astounded at the temerity of its former vassal, imposed a blockade and persuaded major oil firms to boycott Iranian crude. Iran might now control its oil, but it could not sell any. London was motivated by a fading empire' unfortunate mixture of vanity and vulnerability. As Hugh Thomas wrote in the droll fashion of twentieth-century British historians, "Ever since Churchill converted the Navy to the use of oil in 1911, British politicians have seemed to have had a feeling about oil supplies comparable to the fear of castration."

Mossadegh did not give in. In 1952, he severed relations with Britain and closed its embassy in Iran, forcing the withdrawal of Her Majesty's diplomats and spies. The British felt that only America could get rid of Mossadegh, but President Harry Truman, believing the problem wasn't Iranian impertinence but British pride, would not back a coup. The calculus shifted when Dwight Eisenhower became president and his administration fell under the spell of British claims that Mossadegh would let Iran slip into the Soviet orbit. Iran's Cold War importance was hard to exaggerate. In addition to having a lengthy border with the Soviet Union and possessing the world's second-largest reserves of oil, it sat along the Strait of Hormuz, through which much of the Middle East's oil was shipped. Although it was greatly implausible that Mossadegh, a die-hard nationalist, would turn pro-Soviet or join hands with Iran's pro-Communist Tudeh Party, Eisenhower, at the dawn of the Cold War, did not want to risk "losing" Iran. He green-lighted a coup that would be overseen by a CIA agent who was a grandson of Teddy Roosevelt's.

Mossadegh was, in many ways, an easy target. He had an erratic personality, laughing uncontrollably at times, crying and fainting, often communicating in whispers with the foreign envoys he received in his bedroom, where he conducted affairs of state while dressed in pajamas. It was not hard for Kermit Roosevelt Jr., with hundreds of thousands of dollars at his disposal, to bribe military officers, politicians and editors. Most of Tehran's newspapers, which accused Mossadegh of being a Communist and a Jew, came under CIA influence, and most of the anti-Mossadegh protests that preceded the coup were stirred up by these bribed politicians and journalists.

The coup began on the night of August 15, 1953. John le Carré could not have invented a better plot. Roosevelt waited at a CIA safehouse, sipping vodka with colleagues and singing show tunes, including "Luck Be a Lady Tonight." Shah Reza Pahlavi, the timid monarch who was a figurehead but whose status would be vastly enhanced by the coup, waited at a seaside villa. Partial to race cars and nightclubs, the reluctant shah had had to be flattered, bribed and cajoled by Roosevelt before agreeing to join the coup. The plan was for a group of pro-shah soldiers to arrest Mossadegh, and for the shah to take charge of the country.

Instead, the pro-shah soldiers were arrested outside Mossadegh's residence. When dawn broke with news of the failed coup, the shah boarded a twin-engine Beechcraft and fled to Baghdad. Roosevelt, however, did not retreat. Although troops loyal to Mossadegh were deployed throughout Tehran, Roosevelt instructed his agents to organize so-called black crowds to shout support for Mossadegh and communism while beating up bystanders and looting shops. It was a violent smear job. As Roosevelt later wrote, "The more they ravaged the city, the more they angered the great bulk of its inhabitants."

Roosevelt also paid for other crowds to hit the streets—except these were in favor of law and order and the man who, they proclaimed, would bring it to the now-chaotic city: Shah Reza Pahlavi. A $10,000 bribe was even given to a religious leader to bulk up the pro-shah crowds with Islamic devotees. Roosevelt had shown a dark genius. Chants of "Death to Mossadegh!" mingled with "Long live the shah!" As CIA rent-a-crowds seized government buildings, pro-shah troops ransacked Mossadegh's house. With his control of the city lost, the prime minister surrendered. The shah, returning home in triumph, told Roosevelt, "I owe my throne to God, my people, my army—and to you!"

The White House was not foolish enough to raise a "Mission Accomplished" banner, but it was delighted with the ousting of the inconvenient Mossadegh. Yet like the invasion of Iraq a half century later, the intervention led to disaster for America and the Mideast nation whose destiny and oil it hoped to control.

Though on opposing sides of the desert front line in 1990, George H. W. Bush and Saddam Hussein shared a predicament: they could not call the war by its name. Even for a dictator, it is not acceptable to announce that you are going to war to seize another country's oil. Principled excuses must be offered. Saddam spoke of Iraq's long-dismissed claim to Kuwait and portrayed the invasion as restoring territory that colonial mapmakers had lopped away. He also accused Kuwait of steal-

American marines in the burning oilfields of Kuwait during the Gulf War of 1990–91

ing oil by slant-drilling across the border (that is, drilling at an angle rather than straight down). Officially at least, invading Kuwait for its oil and the power and glory it would bring to Iraq was the last thing on Saddam's mind.

Bush was in a bind too, because Americans, horrified when they hear that Kathie Lee Gifford's clothing line is made with child labor, also do not want to hear that the gas in their SUVs requires the shedding of blood. Such truths were unspeakable, literally. This was acted out years before in the brilliant film *Three Days of the Condor*, which starred Robert Redford as a naïve CIA researcher and Cliff Robertson as his cynical boss. Redford, learning of CIA killings related to the Middle East, tells Robertson that Americans will not support murder for petroleum. "Ask 'em when they're running out," Robertson replies archly. "Ask 'em when there's no heat in their homes and they're cold. Ask 'em when their engines stop. Ask 'em when people who have never known hunger start going hungry. You wanna know something? They won't want us to ask 'em. They'll just want us to get it for 'em."

A month after Iraq's invasion of Kuwait, President Bush made his case in a speech to Congress. The war would not be for oil but for the rule of law and the sanctity of human rights. Bush quoted from a letter an American soldier, based in Saudi Arabia, had sent to his parents. "I am proud of my country and its firm stance against inhumane aggression," Private Wade Merritt had written. Bush outlined the steps he would take to "defend civilized values around the world," and he spoke for the first time of the grand design the war would buttress. "Out of these troubled times," he said, "a new world order can emerge. . . . A world quite different from the one we've known. A world where the rule of law supplants the rule of the jungle." He did not mention oil until one-third of the way through his speech, then quickly moved on.

It was true that Iraq's invasion was illegal and brutish, and that if it was allowed to stand, the post–Cold War era would begin with a major violation of international laws. But international laws were violated every day across the globe. Liberia was being savaged in an atrocity-filled conflict that the White House rarely bothered to condemn. The call to high morals was particularly odd because American soldiers would sacrifice their lives in Kuwait to restore an all-powerful monarchy that had disbanded the parliament and banned political parties. And for years the U.S. government had abetted Saddam's regime. It was an odd time and an odd place to champion war for democracy. It didn't add up.

Americans sensed this; they were not persuaded about the need to go to war, according to opinion polls at the time. Something else was needed. A month after Bush's "new world order" speech, a poster child for war emerged. Nayirah, a teenage Kuwaiti girl, testified at a congressional hearing. She tearfully recounted what she'd seen while visiting Al-Adan Hospital before slipping out of occupied Kuwait: "I saw the Iraqi soldiers come into the hospital with guns. They took the babies out of the incubators, took the incubators and left the children to die on the cold floor." Nayirah's last name was not disclosed because, the hearing's organizers said, her relatives trapped in Kuwait would face retribution from the Iraqis.

Nayirah's testimony spread across America in TV broadcasts and newspaper stories. It went viral. An Amnesty International report (later retracted) raised to more than three hundred the number of infants removed from incubators. Bush reinforced the theme, saying during a speech at Pearl Harbor (no less) that dialysis patients were ripped from their machines and twenty-two babies died after being taken from incubators. In the debate before the Senate voted by a slim margin, 52–47, to approve the Gulf War, seven senators cited the incubator story.

There was just one problem. Like the tales of German soldiers bayoneting Belgian babies during World War I, and like the stories of weapons of mass destruction in 2003, it wasn't true.

Nayirah was the daughter of Kuwait's ambassador to the United States and a member of the royal family. Her testimony was arranged by Citizens for a Free Kuwait, which, despite its populist name, was a front created by the royal family to channel more than $10 million to the public relations firm Hill & Knowlton for a campaign in favor of war. Nayirah's story was shown to be a fabrication. The hospital had only a few incubators at the time, and according to the hospital staff Iraqi soldiers did not throw babies onto the floor.

To generate support for the war, President Bush was obliged to obscure the truth from the public, just as his son would feel obliged to massage the facts twelve years later. Privately, President Bush was honest. Before major military or diplomatic moves, American presidents tend to issue national security directives that circulate at the highest levels of government. If you want to know the truth of war and peace, these secret directives are more useful than public speeches, and thanks to the Freedom of Information Act, it is occasionally possible to know their contents. In National Security Directive 54, issued a month before the liberation of Kuwait, President Bush mentioned oil in the first line and never mentioned incubators or democracy. "Access to Persian Gulf oil and the security of key friendly states in the area are vital to U.S. national security," NSD 54 began. "The United States remains committed to defending its vital interests in the region, if necessary through the use of military force, against any power with interests inimical to our own."

On February 24, 1991, a coalition led by a half million American troops crossed into Kuwait and routed Iraqi forces in one hundred hours of ground warfare. As they retreated, Saddam's army set alight more than six hundred wells. The battlefield was bathed in oil.

For Iraqis, the Gulf War of 1990–91 was not an anomaly in the annals of oil and invasion. Osama Kashmoula was thinking of even more distant events when I found him at the Oil Ministry in 2003.

The ministry's leadership, handpicked by Saddam, had gone into hiding, leaving senior technocrats like Kashmoula to do what they could to get things going in the postinvasion vacuum. It had been just a few days since the events of Firdos Square. I heard Kashmoula's voice at the end of a corridor that had the vacant, unpeopled feel of a ghost town. A breeze rustled through windows shattered by the shock waves of bombs. His office door was open and he was shouting into a satellite phone that had been lent to him by an American officer. Kashmoula was a short and thickset man in his fifties, and like most of Iraq's engineers he was fluent in English. It was one of the odd facts of life that outlaw Iraq was home to some of the best engineers in the gulf. Saddam's regime, before achieving its rogue status, had sent its most promising students to British and American universities for postgraduate training, and decades of war and sanctions had turned these men and women into ingenious desperados. They were the MacGyvers of the oil world.

The least surprising thing about Kashmoula was that he was an optimist. Nationality does not matter; all oilmen are optimists. It stems from digging a dozen dry holes before striking oil, and from the tremendous hazards that are regularly overcome to extract the treasured liquid. Oil is beneath layers of Alaskan ice? No problem. It is under treacherous seas off Sakhalin Island? A way will be found to get it. In the middle of a desert? It shall be extracted forthwith. Trained with this mind-set, Kashmoula was hopeful for the future because United Nations sanctions that had strangled Iraq would be lifted and America, with its world-class technology, would open its arms. He recalled visiting Russia a few years earlier and realizing that Russia's best machinery came from Germany. If things worked out right, Iraq would be rebuilt with Western hardware and would produce more oil than ever before and make everyone happy—not just Iraqis, who would become rich, but also Americans, who would have a new friend in the Mideast and cheap gas in their cars.

Yet Kashmoula was bothered by one thing.

"The Americans and British are talking to the Iraqi people as though we are naïve," he said. "You say you are here to free us, but this is ridiculous."

His face assumed a knowing look, somewhere between a scold and a laugh. He mentioned Stanley Maude, the general who led British troops into Baghdad in 1917. Maude is all but unknown in America and little remembered in Britain, but Iraqis memorize his role in their history. Before the end of World War I, Baghdad was a regional capital of the tottering Ottoman Empire. Maude led a British expeditionary force into the city, signaling its involuntary transfer into the custody of the British crown. Maude soon issued his Proclamation of Baghdad, in which he stated, "Our armies do not come into your cities and lands as conquerors or enemies, but as liberators." Later that year, Maude died from cholera, one of the first victims of his own triumph.

It was an odd liberation. The League of Nations soon recognized the newly created Iraq but placed it under a British mandate. The British imposed a Sunni king even though Sunnis were a minority, and this prompted the first Iraqi insurgency of the twentieth century, in which Kurds and Shiites fought their foreign occupiers. Iraq gained quasi independence in 1932 but British troops remained, because among the country's attractions, a massive oil field had been found near Kirkuk. Maude's evocation of the harsh Ottoman period that preceded British intervention—"your forefathers and yourselves have groaned in bondage"—predated by almost a century the 2003 speeches of George W. Bush, who also offered a liberation narrative to justify his occupation.

Kashmoula pointed to the bullet holes in the chairs and walls of his office, the consequence of negligent strafing. "They could have been more surgical," he said wryly. What offended him most was not the invasion's violence or his need to acquire new furniture but the excuse, first evoked by Maude, that foreign incursions were for the sake of freedom and democracy. "We are educated," Kashmoula said with a smile.

"I lived in the West for more than seven years. We know you have certain targets and reasons for coming here. We have oil and you need it. The whole world is built around oil, so let's talk about it honestly."

A few weeks later, I saw Kashmoula again in a conference room at the ministry, which was finally bouncing back to life. With April turning to May, more workers were allowed to return to their desks, though marines frisked everyone who entered and exited, lest they bring in bombs or carry out secrets. The conference room, off the marble-floored lobby, had become the setting for a series of first dates in which Iraqi officials met American experts who represented the occupying power but stressed that they were advisers rather than bosses. The irony (or significance) was lost on no one that the first senior American to meet Kashmoula and other top ministry officials was a former ExxonMobil executive, Gary Vogler, who arrived at the ministry one day with a crew cut, flak jacket and bodyguards. Walking into the conference room, Vogler encountered more than a dozen Iraqis and attacked them with a firm handshake.

"Oh, you're the guy we need to work with," Vogler told one of the Iraqis, pumping his hand enthusiastically.

"Good, good," he told another, moving farther around the table.

"Great, we've got work to do with you," he said to yet one more.

And so he circled the room, issuing superlatives rather than directives. Tea was served. I was nudged out of the room, but Kashmoula had already told me that occupier-to-occupied commands were not necessary. The Americans and Iraqis in the ministry shared the same short-term goal, to get oil flowing again. Whether Iraq would withdraw from OPEC, whether it would privatize its fields, whether American firms would receive preferential deals, whether the invasion was for petroleum rather than democracy—for oilmen, these issues were secondary to the task of repairing damaged wells and pipelines.

Yet something odd was happening. The passing inconvenience of postinvasion looting was turning into a permanent condition. A ministry official, Mohammed Aboush, who was responsible for the northern portion of Iraq's oil industry and who was in the conference room with Vogler, told me afterward that his car had been stolen, and that when he traveled to Kirkuk, a hundred miles away, he took a taxi because he didn't want to risk losing another vehicle. Where oil still flowed, smugglers tapped into pipelines. Where it didn't flow, looters walked off with the pipelines, selling them as scrap metal. Iraq was now in the ironic situation of importing gasoline from Kuwait, yet even so—and here humiliation was piled onto disgrace—people waited for days in dystopian lines stretching for miles to buy some of it. Iraq had joined Nigeria and Iran in the dumbfounded ranks of oil exporters that had to import gasoline. "We had meetings with KBR," Aboush said, referring to the Halliburton subsidiary responsible for helping to patch up the oil industry. "They took notes of our problems and promised to look into it. But they are very bureaucratic. So far, nothing. The chain of command"—Aboush broke into laughter at this—"even Mr. Bush cannot do anything, apparently."

Aboush's office reflected the chaos that continued to linger in the country. His computer and his wall decorations, which had vanished during the fall of the regime, had not been replaced. Before U.N. sanctions had been imposed in the 1990s, Aboush had been responsible for evaluating the bids of foreign companies; he'd realized that the worst were Chinese and Russian ("lousy, incompetent") while the best were American. The French were technologically adept but financially unsavory—they would cheat their grandmothers, Aboush said. Of course, none of this mattered back then, as Saddam awarded contracts for political rather than technological reasons. Things worsened once U.N. sanctions limited Iraq's industrial partners to shady firms willing to operate on the margins of international law. The invasion, Aboush hoped, would change everything. As he wishfully told me, "Instead of giving contracts to people who don't know the difference between oil and fish and chips, we will be deciding between Exxon and Conoco."

Yet Aboush was not enjoying the luxury of such choices. The famous disarray of the occupation—the lack of troops to provide security, the scarcity and cluelessness of reconstruction experts—plagued the oil sector, too. Vogler, the former Exxon executive, did not appear in Baghdad until weeks after U.S. troops, and he arrived without a boss. It wasn't until early May that Philip Carroll, a former chief executive of Shell, was named as the top American oil adviser, and he didn't get to Baghdad until weeks after that. The midlevel American official through whom I tried to arrange interviews with Vogler and Carroll was hobbled by the fact that he shared a computer and satellite phone with several other occupation officials, and he did not have his own vehicle to travel between the ministry and the American headquarters at the Republican Palace, a few miles away. With my own laptop, sat phone and SUV, I was better equipped than a key official in the postoccupation oil hierarchy. This was not how Aboush imagined an invading superpower would manage the world's third-largest reserve of oil.

Aboush, who was as round as he was tall and as warm as he was round, launched into an explanation of the attractions of Iraq's oil. He talked of the vast stretches of desert that had not yet been probed, he noted the attractive shallow depths of the nation's reservoirs, he emphasized the low cost of extraction, he praised the prowess of American firms—"even without arm twisting, they will get the contracts"— and he said world prices would fall once oil flooded out of rebuilt Iraq. Why did he bother mentioning the obvious? It struck me that Aboush was urging more American attention to Iraq's oil rather than less. His main fear wasn't that Iraq's petroleum would be owned by Halliburton but that it would be ignored by Bush. An expert in what happens aboveground as well as below, Aboush sensed that Iraq's geological potential might be defeated by its geopolitical dilemma. "We are one of the wealthiest countries in the world, but our wealth has not been used well," he noted. "Oil has not been a blessing. Without oil, we would not have had these wars. But I am a patriot. I want to see my oil spent on my people."

Aboush knew that when America has a monocular aim, it tends to get the job done. After the Gulf War of 1990–91, hundreds of fires in Kuwait burned through millions of barrels of oil a day, but by year's end the fires were out and the oil flowed again. Aboush needed only to make a short and dodgy drive to see that Iraq's oil sector—supposedly the grand prize of invasion—was starved of attention from its occupiers.

· · ·

The bridge across the Tigris to Dora was a double-decker affair with four lanes on each level. Its notability arose not from its design, which was a triumph of concrete over aesthetics, but from its linking of Baghdad to a neighborhood that hosted one of Iraq's three refineries. Dora's refinery, an industrial crown jewel, was targeted by American bombs during the Gulf War of 1990–91. Afterward, it was rebuilt with around-the-clock work that concluded with a visit by Saddam Hussein, who congratulated his team of reconstruction wizards.

A dozen years later, I drove over the bridge to meet Dathar Khashab, the chief wizard. He was a raspy engineer to whom the intervening years had been good—he had become the refinery's director, presiding over a curious testament to the strange American role in the region. More than half a century ago, the refinery had been built by Kellogg Brown & Root and Foster Wheeler, Ltd., which meant that Americans had built a facility they would later bomb. Despite the wallop of precision missiles, the refinery still used, in 2003, much of its half-century-old equipment, including a vintage IBM clock that workers continued to punch their time cards into. The refinery's administrative building, fronted by a grass lawn and semicircle of palm trees, had a winding iron stairwell that was a mixture of Bauhaus and Tara. The sign for the men's bathroom said, quaintly, "Gents." An executive conference room was decorated with a portrait of the refinery's first director, George Mitchell, who served, as a gold-plated plaque attested, from May 25, 1955, until December 2, 1959. Dathar Khashab looked forward to working once again with the foreign engineers who'd taught him and his predecessors everything they knew. He pointed to the portraits of the dozen or so Iraqis who'd followed in Mitchell's footsteps. "This man was a British graduate," Khashab said. "This one, an American graduate. This one, British. This one is living in London. Here, a British graduate." Khashab touched his own chest. "I got my degree at Sheffield University."

He was a chip off the block, shaking hands and slapping backs and barking orders like a profane Texas oilman. He had a KBR construction hat behind his desk and all but worshipped W. Edwards Deming, a 1950s management guru who'd emphasized the role of quality as a route to profitability. "He is a fantastic man," Khashab said. "His book is, for me, like a god." Though a member of the Ba'ath Party—Iraq's business class had to join, Khashab said, defensively—he had always admired the business acumen of the nation that had just invaded his own. The KBR hat behind his desk was displayed without irony.

"It is much easier to work with British and Americans," he added. "We speak the same language, the same standards and equipment. When I talk with Russians and Chinese, it's all obscure to me. I don't know their standards or goals. We have a hard time just getting them to understand what we are talking about."

His Ba'athist affiliation was trumped by his devotion to the steel pipes and pressurized stacks that were his flammable love. Yet he was stunned to realize that although his refinery was crucial to Iraq's oil infrastructure, its survival was not guaranteed. Dora was already on its way to becoming a lawless satellite. In the weeks after the invasion, the short hop from Baghdad to Dora had become a lethal test, because trying to cross the bridge had turned into a carnival of carjacking. Thieves were in the habit of setting up roadblocks at its midpoint or driving alongside cars and aiming guns at drivers to persuade them to stop. The best defense was to wait on the Baghdad side for other vehicles to go first. If bandits attacked the car ahead, there was time to turn around.

Unlike the Oil Ministry yet like the bridge, the refinery did not get prompt attention from the American military. With law and order breaking down even before the invasion's culmination, looters besieged the refinery, which was a paradise for them. Not only did the complex contain an abundance of oil, gasoline and heating fuel, but its accessories included cars, trucks, buses, computers, fax machines, tools, desks and chairs. Even its trees were valuable, because they could be uprooted and sold on the black market. When the attacks began, Khashab scrounged up a hundred assault rifles and handed them out to workers whom he divided up into defense teams. For three sleepless days and nights, Khashab's impromptu militia held looters at bay. The arrival of the marines in Firdos Square did not change things; looters continued to try to sneak and shoot their way into the refinery. And why shouldn't they? The invaders were standing aside as "stuff"—

Rumsfeld's euphemism for anarchy—happened. Another day passed
and the looting siege worsened. The next day, looters issued an ultima-
tum to Khashab: surrender or large-caliber weapons would be used
against him. Khashab put a machine gun at the entrance and manned it
himself, though this was a bluff because there was almost no ammuni-
tion left. He was saved by the arrival of soldiers from the 101st Air-
borne Division. They came only to check the condition of the refinery,
but as Khashab shook the commander's hand he said, "Now that you
are here, you have to provide security."

The incompetence of the American occupation has been amply
documented, but the initial neglect of Dora seemed to reflect more
than the astonishing sweep of American blundering. If oil was the
motive of invasion, surely Baghdad's only refinery would be protected
at the first possibility and afterward provided with all the soldiers,
experts and supplies it needed. The White House might not care about
the National Museum or the Al-Kindi Hospital, but surely the oilmen-
turned-politicians in Washington would look beyond the ministry
building to devote more than passing attention to the preservation of
the petroleum infrastructure. They didn't, and I wanted to understand
why, so I hung around the refinery to see what would happen.

As April turned to May, Khashab found himself in the strange position of asking not for Americans to leave his refinery but for more of
them to come. In May, security duties were placed in the hands of Captain Tom Hough, who, along with fifty GIs, made a barracks out of a
home just inside the refinery's grounds. It was like a frat house in which
corridors were littered not with kegs of beer but M-16s and ammuni-
tion. Their job consisted of turning back would-be looters in a catch-
and-release program. The usual punishment for teenagers was a round
of push-ups before being sent home. Adults had it somewhat worse,
turned loose with their hands bound behind their backs by a shaming
set of plastic cuffs. It was endless work. "We have very efficient
thieves," Khashab noted. "Better than Al Capone."

Khashab and Hough were bound together by isolation. A farm boy
whose hobbies included fishing and hunting, Hough's experience in
the oil business consisted of filling up his car in the Midwest. The

Coalition Provisional Authority (CPA), which was supposed to run the occupation, had not sent anyone to offer him advice or assistance, so he had to cope with a bewildering set of tasks. Workers who hadn't been paid for months demanded money that wasn't there; mysterious banners urged a struggle against the infidels; families with domestic disputes sought his arbitration because the local police and courts were AWOL; workers tried to beat up corrupt managers who had stolen their wages or generally oppressed them; and an Iraqi interpreter was found dead one morning with a bullet in the back of his head and a note on his body that said, "This is what we do to traitors." It was Hough's new job to take care of all of this.

"I don't have any idea what the Bush policy is," he told me. "I don't know what they're planning for the future of Iraq. I am just making it up as we go along because I sure didn't read the latest State Department policy paper."

The pairing was particularly odd because Khashab, the civilian, wanted looters shot, while Hough, the soldier, refused. In these early days of occupation, his unit's rules of engagement allowed lethal force only if soldiers were fired on, and most looters were smart enough not to shoot at the Americans. Khashab thought this was ludicrous, because calisthenics were a feeble deterrent. I sat through dozens of meetings and, in its Alice-in-Wonderland way, the scenario rarely varied. Khashab chain-smoked Gauloise cigarettes while Hough spit tobacco juice into an empty water bottle. Hough, the blond, can-do captain, was dressed in battle fatigues and desert boots, while Khashab, old enough to be his father, wore soiled overalls and workaday loafers. They met in Khashab's office, under a painting that showed the refinery against an orange sky.

"Your efforts will go down the drain if you don't do anything about security," Khashab advised one day. "I mean, you are occupying the country. Secure it! Act more violently!"

"We're not doing a good job," Hough said. "It's embarrassing to me. I don't know how to rebuild countries. I don't know what I'm doing. If we come to a country and destroy its government and destroy its army, we have to rebuild it. But I'm wondering, Where are the people who rebuild countries? I just jump out of planes and kill people."

On good days in the spring of 2003, the refinery limped along at one-third its approximately 150,000-barrel-a-day capacity. American generals and even the occasional CPA official began visiting as May turned to June, drinking tea in Khashab's office, listening to his complaints and, when it was time to go to their next exercise in occupational futility, offering a reassuring pat on the shoulder for a job well done and for the good times that were certainly ahead. Even Paul Bremer, head of the CPA, eventually stopped by and marveled at the refinery's control room, saying that the vintage machinery reminded him of the Flash Gordon television show he saw as a child. But the wizard of America's muddled occupation had little to offer other than words of encouragement. The Dora refinery was like a neglected stray looking for an owner.

Iraq's oil and America's desire for it continued to confound me.

The war-for-oil argument was alluring. The argument tended to be prefaced with reference to Cheney's work, before he became vice president, as chief executive of Halliburton, whose KBR subsidiary was rewarded with billions of dollars in contracts in Iraq during the occupation. The argument referenced Cheney's chairing, in 2001, of a secret task force of government officials and oil executives who discussed energy policy. Bush's background as a Texas oilman, though a failed one bailed out by friends, was evoked in the same breath. Oilmen always desire more oil, so the motive for war seemed as obvious as the tattooed marines in front of the Oil Ministry. Clearly, the Texans in the White House wanted to control Iraq's 115 billion barrels of crude.

But there was a flip side. Oilmen prefer to deal with dictators rather than kill them. In Equatorial Guinea, Saudi Arabia, Kazakhstan, Angola and elsewhere, Americans have not hesitated to nuzzle and sign contracts with tyrants who reside at the top of the repulsiveness scale. History shows this tendency to hold true even in Iraq, where Saddam Hussein's gassing of Iranians and his own people did not harm his sales to America, which held steady until 1990. Days after the invasion of Kuwait, United Nations sanctions halted Iraqi oil exports, but the First Gulf War was only a temporary obstruction.

The sanctions were loosened in 1996 so Iraq could export oil and use the proceeds to buy humanitarian goods. This was the oil-for-food program, and one of its twists was that the largest purchaser of Iraqi oil was, once again, the United States. Even though American F-16s were dropping bombs on Iraq to enforce the no-fly zones in the north and south of the country, American consumers were gassing up with Iraqi crude. Another twist involved Saddam selling discounted oil to buyers willing to pay kickbacks. Chevron was one of the buyers; in 2007 the firm admitted that a trader had funneled money to Iraqi officials. War, sanctions, genocide, laws—these were no more difficult to get over than speed bumps.

American oilmen certainly wanted greater access to Iraq's under-produced fields, but their preference was to lift sanctions that prevented them from working there. That was also their preference in Libya, where Muammar Qaddafi's misbehaving but oil-rich regime had been under sanctions since the Reagan era. Their slogan tended to be "Make business, not war." This was pointed out a year after the invasion of Iraq by Anthony Sampson, who wrote a classic history of political manipulations by oil firms. "It is tempting to depict the war in Iraq as a straightforward attempt to control its oil reserves in the interests of the big corporations," he noted in 2004, just months before he died. "[But] both Sir Philip Watts of Shell and Lord Browne of BP were warning that war in Iraq was likely to destabilize supplies and antagonize other Islamic oil producers. It might seem surprising, but it has happened before. When Sir Anthony Eden launched the Suez War in 1956, also claiming to defend British interests, he did not consult Shell or BP, which had the most to lose. Both companies were deeply worried that such a dangerous adventure would antagonize Arab oil producers throughout the Middle East—which it did." Sampson concluded, "Governments that are bent on military adventures—contrary to most conspiracy theories—become curiously resistant to advice from commercial concerns, which often understand much more about the consequences." The Gulf War of 1990–91 offered an example of the nonre-wards of military action in the modern Middle East. After their American-assisted return to power, Kuwait's royal family did not let its gratitude extend to the granting of oil deals to American firms; minor reconstruction contracts were doled out to them but not the invaluable production-sharing agreements that are the manna of the oil industry.

The Oil Ministry was just one of many places in Iraq and America where people had told me, "It's all about oil." Literally, the phrase *was* true. The keyboard I type on, the clothes I wear, the heat in my apart-ment, the shoes on my feet, the pots in my kitchen, the cars and planes that transport me—they are oil's spawn. When I visited an oil museum in Saudi Arabia, an exhibit noted that almost everything in the indus-trial world contains oil—plastics, fertilizers, even toothpaste. Products that don't contain oil depend on it to be manufactured and moved to markets: the wood in my desk was almost certainly felled by a gas-fueled chain saw and transported to me by a ship, train or truck that would not budge an inch without fuel for its engine.

The question is not *whether* war is about oil but *how* it is about oil. Even Paul Wolfowitz, the neoconservative deputy secretary of defense and the official with the highest quotient of pro-democracy idealism in the Bush administration, had oil on his mind. His interest in Iraq, as James Mann wrote in *Rise of the Vulcans*, dated from the 1970s and stemmed from his concern that a regime hostile to U.S. interests would dominate the oil fields of the gulf. There are complexities behind the everything-for-oil philosophy. Even psychological issues might have played a minor role in the decision to invade, because as Bush famously said of Saddam, "This is the guy that tried to kill my dad."

In the 1970s, in *Shah of Shahs*, Ryszard Kapuściński described oil as a substance that "anesthetizes thought, blurs vision, corrupts." His data point was petroleum-mad Iran, but his thesis applied broadly. Amer-ica's desires were so influenced by Iraq's inebriating crude that Wash-ington could not think straight about the reasons for invading. And it was a difficult time to think straight. The post-9/11 political climate in America was characterized by enormous fear of another terrorist attack. The specter of dirty bombs caused more concern in Washing-

ton than $4-a-gallon gasoline. The WMD threat from Iraq has proved
to be nonexistent, with the "evidence" having been concocted or
manipulated, but that does not mean Bush and Cheney did not believe
in this peril that didn't exist. A threat need not be real for it to inspire
fear or exaggeration.

The chore of figuring out oil's role in the 2003 invasion has been
complicated by the Bush team's reluctance to discuss the subject. I prefer lies to silence, because lies yield information. Rumsfeld's insistence
that oil had literally nothing to do with the war was unintentionally
revealing, in the fashion of "the lady doth protest too much." Yet clues
were provided, notably by the invasion's most notorious and notori-
ously silent instigator.

On August 26, 2002, Cheney visited Nashville and spoke to a convention of the Veterans of Foreign Wars. He repeated the charge, later
proved false, that Iraq had resumed its efforts to acquire nuclear
weapons. His speech came to a crucial passage that illustrated, in its
description of Saddam's alleged menace, how oil was deeply embedded
with other motives for war.

"Should all his ambitions be realized," Cheney said,

the implications would be enormous for the Middle East, for
the United States, and for the peace of the world. The whole
range of weapons of mass destruction then would rest in the
hands of a dictator who has already shown his willingness to
use such weapons, and has done so, both in his war with Iran
and against his own people. Armed with an arsenal of these
weapons of terror, and seated atop ten percent of the world's oil
reserves, Saddam Hussein could then be expected to seek dom-
ination of the entire Middle East, take control of a great portion of the world's energy supplies, directly threaten America's
friends throughout the region, and subject the United States or
any other nation to nuclear blackmail.

In his eight years as vice president, Cheney was cynical, manipula-
tive and inept. History might judge him the most ruinous public ser-

vant ever. But it is hard to conclude from the available evidence that he did not believe at least a portion of his fear-mongering about WMD. The evidence suggests that oil and WMD reinforced each other, like volatile elements in a laboratory that explode when combined. And it's important to remember that Cheney was not commander in chief. President Bush had his own mixture of priorities and fears, as did the men and women whom he listened to. Neither Cheney's motives nor the motives of the administration he served can be distilled into one word. WMD, democracy, religion, Oedipus, oil—America was like a drunk fumbling with a set of keys at night.

What does our desire do to others? In its impact on Iraq, oil was like a volcano that erupted every decade or two with a new cataclysm. Because it can be a hindrance to stand atop the volcano as it erupts, I left Baghdad and found a calmer station a hundred miles away in Najaf, where I could get a different perspective.

Najaf is home to the tomb of Imam Ali, whom the world's Shiites regard as the successor to the prophet Mohammed. It is a holy city, and one of the greatest Shiite seminaries is located there, along with some of the faith's greatest thinkers and leaders. Najaf attracts a stream of pilgrims who visit the gold-domed mosque that contains Imam Ali's remains, and many of these worshippers bring their dead for burial outside the city. The caskets are carried around the mosque and then taken to Wadi al-Salam, the Valley of Peace, where millions of people were buried.

Even though Najaf had more dead souls than barrels of crude—the largest reservoirs were hundreds of miles away—it had much to teach about oil's impact. One of the best classrooms was a small, ramshackle building on an alley near the mosque. I visited several times, and people were always pushing, shouting and pleading to get inside; the guards, who were hard men, let in just a few at a time. After taking off their shoes, the visitors entered a darkened atrium that was like a musty library that had not been dusted for years. It was crowded with clerics and religious students who talked in whispers and smoked Yemeni cigarettes. (American products were *haram*, forbidden.) Overhead fans

A militiaman's AK-47 in front of
a poster of Muqtada al-Sadr

creaked and kept the air moving but not fresh or cool. One day I was taken to a small, windowless upstairs room that was furnished with a long-suffering couch and a similarly distressed armchair. I sat on the couch and waited. The religious leader Muqtada al-Sadr would arrive soon, I was told.

There was time to ponder what Iraq's oil reserves had paid for in the quarter century of Saddam's rule. It underwrote roads and universities and hospitals, but its nonconcrete legacy was disastrous. Since the 1960s, there had been repression, war, poverty, more war, more repression, more war. Oil creates a particularly volatile type of grievance. It is one thing for people to be poor due to factors beyond a listless government's control—being landlocked or drought-stricken, for example. It is another thing for poverty to proliferate under a hated regime that plunders or wastes an immense amount of wealth. On the eve of the American invasion, with government institutions discredited and inept, religious figures possessed the preponderance of authority in Iraq. Shiite and Sunni clerics were the answer to the question of where people go for succor, voice and direction if their government fails them for decades and, in the wake of invasion by an army of infidels, ceases to function.

When the Nigerian state began to collapse on itself in the 1980s, there was little to take its place or shore it up because the country was fractured along hundreds of ethnic and religious lines; low-level anarchy ensued. In Najaf, the collapse of Saddam's petrostate had created the spectacle of people literally crawling to seek help from religious figures like Muqtada al-Sadr; in Sadr's office one day, I saw petitioners prostate themselves in front of him, asking for assistance. Sadr was about thirty years old and still a student of the Koran, but his power derived from his revered father, an outspoken grand ayatollah who was murdered in 1999 by presumed Mukhabarat agents.

The violence that would consume Iraq in its post-Saddam era had already begun in Najaf. As Saddam's regime fell apart, an exiled Shiite leader, Abdul Majid al-Khoei, was flown into the town by the American military, which hoped to insert a pro-Western voice. When Khoei made his first visit to the tomb of Imam Ali, a crowd chanting Sadr's name beat, stabbed and shot Khoei to death. One version of events had the crowd dragging him to Sadr's office a few hundred yards away, where Sadr gave the order for execution, because Khoei's family was a rival to his own. Among the Shiites, who were the long-repressed majority in a country dominated by its Sunni minority, sorting out the future would not be a peaceful affair. This, too, was a legacy of oil, as I was about to be told.

Sadr swept into the room with several aides, most as young as he was. Dressed in a black gown and turban, he moved at a forward angle, leaning ahead, as though he was in a hurry or lost in thought or both. There was no shaking of hands, no exchange of pleasantries, no eye contact as he sat in the armchair kept vacant for him. It was as though a dark cloud had moved into the room and sat before me. He nodded at my interpreter in a way that meant, *Begin.* He showed no emotion aside from graveness, as I had seen him do the day before, while delivering a sermon at the Kufa mosque, where his assassinated father had delivered his sermons. In addition to killing his father, Saddam's regime had murdered Muqtada's brothers and a famous uncle, whose body was set on fire after nails had been drilled into his head.

Sadr had mentioned, at his sermon the day before, that enemies would try to stand in the way of Iraq. I asked whether he had Americans in mind. He didn't hesitate to say that President Bush wanted to carve up and weaken Iraq, the better to control its oil. "Everyone knows that America is not looking for reforms to unify the country," Sadr told me. "They will be an enemy to us or, shall we say, they will not be a friend to us. We are looking for a unified Islamic nation, so we think our aim is different than their aim." This theme became explicit in the years ahead, as crowds carrying Sadr's portrait marched in Najaf and shouted, "Oh, occupier, our oil is for us, not you."

Violence took its inevitable course, as Sunni and Shiite militias fought the American occupiers and fought each other. For some Iraqis who took up arms, oil was not just a nationalist rallying cry but a source of funding for their efforts. About $200 million a year in smuggled oil went into the coffers of militias, according to a U.S. government report, though the actual figure was probably higher. This was done by tapping oil from pipelines (as rebellion-for-profit militias did in Nigeria) or by demanding a cut of the revenues earned by the state oil company. Until the Iraqi army threw it out in 2008, Sadr's Mehdi militia was in control of Basra, where it received a percentage of the oil that was exported from the local port. Up north, Kurds and Arabs who lived in Kirkuk claimed rights to the giant oil field outside the city; the standoff might well be resolved by force. Even as the anti-American insurgency calmed down in 2009, a new round of oil warfare beckoned.

Just as America had other concerns when it invaded, Iraqis killed each other for reasons distinct from petroleum. But without a doubt, oil was a violence-inducing intoxicant for the people who lived atop it as well as the foreigners who desired it. The country has too much for its own good. It seems reasonable to consider that the fighting may not stop until the wells run dry.

8

Alienation

Ali al-Naimi is barely five feet tall but can be found in a crowd quite easily, because he is always at its center. Naimi, as you'll recall, is the minister of oil for the Kingdom of Saudi Arabia, so the center is wherever he places his Guccis. On a warm spring evening in Washington, D.C., he was the easy-to-locate guest of honor at a reception at the Ritz-Carlton hotel, where his welcomers included envoys from America's financial behemoths. Merrill Lynch sent a representative from its executive suite, as did Morgan Stanley, Exxon, ConocoPhillips, Dyncorp, General Dynamics, U.S. Steel, Chevron, the World Bank, the International Monetary Fund, the Department of State, the Department of the Treasury and the Department of Energy. Almost everyone wore pins of side-by-side Saudi and American flags that were handed out at the door. But it was an evening in which desire mixed with anxiety, like a nineteenth-century ball at which a hundred suitors make flattering compliments to just one belle.

"We're trying to get some of that Saudi oil," I heard one businessman confide to another.

"It's hard to get," the other replied, with adolescent envy.

Naimi swept into the ballroom a half hour late, his slight runner's frame giving him a birdlike demeanor. He moved as quickly as the central banker of oil might be allowed to move in such a crowd. Hands were extended and shaken, warm greetings were made, laughs were plentiful even if the joke was not heard and, in general, confidence and confidences were encouraged. Even the highest-ranking executives, feigning incidental interest, tilted their heads to eavesdrop as Naimi hopped past. They listened to his brief speech as though their destiny was being revealed. "We are the biggest exporter of crude oil and the U.S. is the biggest consumer of crude oil," Naimi began. "That makes for excellent complementarity." He smiled, cueing polite laughter.

Naimi was born in 1935 in the desert around Khobar, a small port on the Persian Gulf. He was as unprepared for modernity as Saudi Arabia itself. His parents were Bedouin, and the family migrated endlessly with their sheep and camels. When he began tending his tribe's livestock, he was told not to wander out of sight of their tents, lest he get lost in the infinite desert. His life was destined to be hard, because well into the twentieth century, the Bedouin suffered the same deprivations as their ancestors centuries earlier. But Naimi had the good fortune to be born around the same time as American geologists began looking for oil in the Saudi desert not far from where his tribe wandered. The discovery of oil showered Saudi Arabia with money, paying for highways and palaces and turning a boy nomad into a globe-trotting minis-

The Kingdom Center in Riyadh

ter whose mere words could alter the world's economy. Oil remade the country, but it could not produce tranquillity.

A country's birth is rarely a peaceful event. It's often the result of violence, and Saudi Arabia was no exception.

The country's founder, Abdul Aziz bin Abdul Rahman bin Faisal al-Saud, after whom it is named, unified his kingdom in 1932, acquiring new territory through military victories and strategic marriages into rival tribes (he had at least seventeen wives and a number of concubines). But within a few years his project was floundering. The king borrowed money from almost every business with funds to spare, and still civil servants were not paid on time. In the Middle East, a king without money is not king for long, so to keep his household and his nation afloat, Ibn Saud, as he is known to westerners, had to take out loans from his personal banker. He finally arranged a more secure lifeline from a distant source—Standard Oil of California, which paid £55,000 in gold for exploration rights. As historian Madawi al-Rasheed has noted, "The oil concession came at a time when the state lurched from one financial crisis to another . . . [and it] resulted in immediate relief."

The exploratory team, led by a geologist named Max Steineke, set up a crude camp near Khobar, sleeping in tents by night and occasionally riding camels by day to survey the desert. The color of their skin was new to the Bedouin, as was the behavior of these interlopers, who started the day not with prayers but, sometimes, with calisthenics. Their searches paid off. In 1938, at an exploratory well known as Dammam 7, the Americans pierced a vast reservoir. After news of the discovery was cabled back to San Francisco, Steineke's camp was augmented with rudimentary air-conditioning and other amenities that meant the Americans would be staying for a while. A year later, Ibn Saud opened a valve that let the first Saudi oil flow onto a tanker ship, the *D.G. Scofield*. The future of the country and the planet shifted at that moment, yet few people noticed. The American government did not even have an embassy in Saudi Arabia—its nearest diplomat was in Egypt. The only Americans at the ceremony were oilmen.

Saudi Arabia was, at the time, one of the poorest nations in the world. Largely illiterate and preindustrial, it had meager exports and minimal relations with the outside world. Ibn Saud rarely ventured outside the country, and hardly any of his few million subjects had done so, except in seasonal migrations with their livestock. The royal court was delighted with its new source of revenue, but there were doubts among the nation's conservative population, always wary of outsiders. How would their lives be changed by oil? Who would truly benefit from it? One of the best portrayals of this dawn-of-oil era is found in the novels of Abdelrahman Munif, especially *Cities of Salt*, which begins in an oasis town whose residents do not understand what the white-skinned visitors are looking for. A skeptical tribesman, wary of foreigners who promise wealth for everyone, warns his friends, "What does it concern them if we get rich or stay just as we are? Watch their eyes, watch what they do and say. They're devils, no one can trust them."

Through the 1950s and 1960s, the American-Saudi relationship remained cozy and sleepy. The extraction of oil rose gradually, under the control of Aramco, a consortium of American companies led by the very fortunate Standard Oil of California (now known as Chevron). With oil costing two or three dollars a barrel, Saudi Arabia had a steady but not extravagant stream of revenue. Its oil was not even needed by the United States, which was, until the 1950s, a net *exporter* of oil. But as America's economy expanded in the 1960s, along with its oil-intensive culture of cars and suburbs, the once-vigorous fields of Texas, Oklahoma and Louisiana were being depleted. By the early 1970s, American production had peaked and the country was importing one-third of the petroleum it consumed. The little-noted shift that had begun when Ibn Saud loosed the first shipment of Saudi oil onto world markets was now going to be felt by the entire world.

The Saudis and their allies in OPEC realized that their customers could be made to pay far more for the oil their economies were addicted to. In 1973, the colonial era of oil, in which American and European companies controlled pricing and distribution, came to an end. When the Yom Kippur War broke out in October and America provided weapons to Israel, Arab members of OPEC boycotted shipments to

America. Prices soared instantly. The reversal was celebrated by Sheikh Ahmed Zaki Yamani, the Saudi oil minister at the time. "The moment has come," he exulted. "We are masters of our own commodity."

This was only partly true. The royal family might be the master of the country's oil, but not of the alienation it fomented.

In today's agricultural and industrial economies, the government does not tend to own every ear of corn that comes from a farm or every slab of steel that rolls off a factory line. Yes, some land and factories might be state-owned, but even in those cases there are usually large numbers of workers and managers involved in the creation of this wealth who care about what happens to the products they have made, and whose salaries come out of the revenues gained by selling them. Other inputs must be paid for—fertilizer, seeds, furnaces, delivery trucks and so on. That is why diverse economies with labor-intensive sectors (whether agricultural or industrial) are healthier over the long term than ones that depend on a single resource or industry: they generate jobs, they are not tossed up or down by inevitable price swings for a particular commodity or product, and the national treasury is not a bulging piggy bank from which princes can take as they wish.

Oil offers another paradigm. In most countries, oil is not owned by the person who lives atop it. (The United States is one of the handful of countries where property rights extend below the earth.) And unlike the farming of wheat or manufacturing of cars, relatively little labor is required to get oil out of the ground, particularly in the Middle East, where the reservoirs tend to be large and close to the surface. This means that an enormous amount of money that flows into the state's hands has no owner or benefactor other than the state. A king or prince who pilfers tax funds is taking cash from his people's pockets, and they will notice. Oil revenues are not squeezed from individuals, as taxes are, and the financial complexity of oil contracts and oil sales allows ample opportunities for theft. The scholars Terry Lynn Karl and Ian Gary noted in a coauthored study that "petrodollars actually sever the very link between people and their government that is the essence of popular control." This is an origin of the House of Saud's moral downfall.

Before the oil rush, the monarchy was known for piety and thrift under Ibn Saud, who lived in a modest palace made of mud bricks. He was not in it for the money or the easy living. He died in 1953 and was succeeded by a son whose immodest ways included throwing money from the window of his royal sedan and watching as his destitute subjects grabbed the riyals fluttering in the air. He was nudged aside by siblings concerned that the monarchy would be destroyed by his rule. But the genie was well out of the bottle. The heirs of Ibn Saud, who had at least forty-five sons and some two hundred daughters, allotted themselves sizable allowances from the nation's oil revenues. The phrase "Saudi prince" became a synonym for fantastic riches. Long before Hollywood A-listers could afford private jets, Saudi royals rented out entire floors of luxury hotels in Switzerland and arrived in silver-plated Boeings; King Fahd, who ruled from 1982 to 2005, reportedly had a fountain in his 747. Saudi royalty all but created the spectacle of "the entourage."

There were two ways of drilling into the public till to fund lavish lifestyles. First, there were direct allowances to royals; the amounts have never been divulged but are regarded as substantial, because palaces cannot be built cheaply. Princes who wanted even more cash used their royal status to win contracts and commissions on government deals. For example, a prince who was a longtime defense minister was famous for the cuts he demanded from foreign companies that wished to sell weapons to the kingdom. Decades ago, an elegant portrait of these inelegant affairs was crafted by William Morris, Britain's ambassador to Saudi Arabia. In cables that were unsealed just a few years ago, Morris described Saudi commerce as "a jungle inhabited by beasts of prey in which one must move with caution and uncertainty." He wrote in another dispatch, "What they do with the wealth is often comedy and sometimes farce . . . the theatre of the absurd is never far away. . . . The sheer effrontery is breathtaking of a prince who will keep on talking about rights and wrongs when you know (and he probably knows that you know) that his cut may be 20 percent of the contract price."

The theater of the brazen was nearer. Prince Bandar bin Sultan, a longtime Saudi ambassador to the United States, said in a television interview that it wasn't a problem if the elite "misused or got corrupted with $50 billion," because a far greater amount of money was not stolen. The prince, whose vacation compound in Aspen was valued at $135 million when it was put up for sale, famously added, "What I'm trying to tell you is, So what? We did not invent corruption." A former fighter pilot, Prince Bandar was himself accused of receiving one billion pounds from BAE Systems, the British arms manufacturer that sold $80 billion of weapons to the Saudi government. An initial probe by the British government's Serious Fraud Office was shut down by then–prime minister Tony Blair because it imperiled British-Saudi ties, though the U.S. Department of Justice has begun looking into the case.

Royal corruption was not the only precursor to instability. The extraction of oil required the construction of an infrastructure that included drilling rigs, pipelines, roads, power plants, offices and airports. American firms flooded into an isolated nation that, until the 1930s, had almost no paved roads. The oil boom functioned as a sudden introduction to Western ways. Elvis Presley and Johnnie Walker were brought to the kingdom, as was education for women. The transition was uneasy because Saudi Arabia was one of the most conservative societies in the Arab world, with a brand of Islam, known as Wahhabism, that was famously strict and puritanical. Even the advent of TV was an occasion for a violent protest from conservatives opposed to reproductions of the human image. While it was a comfort to have air-conditioning in Riyadh and running water in Jeddah, some Saudis worried that Islam's holy land, where Mecca and Medina are located, was selling its soul.

The revolt of the alienated began in 1979, when several hundred gunmen, led by a messianic fundamentalist named Juhayman al-Oteibi, seized Mecca's Grand Mosque, Islam's holiest shrine. Perhaps the seizure would have taken place without the oil wealth on which the royal family had compromised itself, but the qualities that made the House of Saud so odious to Islamists—its Westernized ways, its love of material comforts, its less-than-devout inclinations, its willingness to allow infidels and their impure customs into the holy land—were linked to petroleum. Oteibi, speaking over a loudspeaker system normally used to broadcast prayers, called for the ousting of the royal family and the imposition of theocratic rule.

The national guard failed to retake the mosque in a series of assaults. The ruling family, realizing that its own forces were inept or disloyal, called upon the military services of foreign countries, including Pakistan and, controversially, France. Infidels are not allowed in Mecca, so the use of French commandos has been officially denied. The foreigners, according to some news accounts, were required to make quick conversions to Islam before entering Mecca. In the final battle, which killed hundreds, a gas was used to disable the fundamentalists. More than fifty-five of them, including Oteibi, were captured and, a month later, beheaded in the largest mass execution in Saudi history.

The royal family felt too weak to oppose a rising tide that its misbehavior had helped incite. Instead, the House of Saud decided to placate the fundamentalists. New authority was granted to religious leaders, with the Ministry of Education placed under their control and enhanced powers given to the *muttawa*, an Islamic police force whose official name was the Committee for the Propagation of Virtue and the Prevention of Vice. To augment its conservative credentials, the government gave several hundred million dollars a year to the mujahideen who fought the Soviet army in Afghanistan. Most crucially, in terms of the global spread of fundamentalism, billions of dollars were poured into projects to propagate Islamic values. An American think tank estimated that more than $70 billion was spent between 1975 and 2002 on projects that included the construction and operation of mosques and study centers. Thanks to this funding, in Karachi, Jakarta, Hamburg and elsewhere, young Muslims were taught that the Koran should be the law of the land, that living as the Prophet lived was a way forward, and that nothing was more glorious than dying in the service of Allah. With a population equal to only 1 percent of Muslims worldwide, Saudi Arabia funded more than 90 percent of the faith's costs, and these expenditures constituted the most expensive information campaign ever mounted.

Paradoxically, the dissatisfaction the monarchy sought to cure by supporting fundamentalist causes had the effect of deepening the problem. Exhibit A is the seventeenth child of a onetime bricklayer.

The oil boom enriched a number of nonroyal Saudis, including Mohammed bin Laden, who was a manual laborer for Aramco in the 1940s. Mohammed used his talent with numbers—he could instantly solve calculations—as well as his proclivity for hard work and harder bargains to create a construction firm that by the 1960s was building everything—roads, palaces, mosques. The wealthiest nonroyal in Saudi Arabia, he was rich enough to support twenty-two wives and fifty-four children. If the oil boom had not happened, he could not have dreamed of such an extended family, and Osama would not have been born.

After Mohammed bin Laden died, in a 1968 plane crash, his firm continued to prosper under the leadership of his eldest sons. At the time of the 9/11 attacks, the Saudi Binladen Group, as it was called, employed about thirty-six thousand people. Osama was not an active manager, though he performed minor functions during quieter moments in his life. Although his wealth was estimated in some reports to be as high as $300 million, later and more reliable estimates cited in Steve Coll's *The Bin Ladens* suggest that Osama received about $24 million from the family firm between 1970 and 1993 or 1994.

This was sufficient to make him, when he went to fight the Soviets in Afghanistan in the 1980s, a noted figure of the anti-Communist jihad who leveraged his personal finances by raising donations from wealthy friends and acquaintances. By the time he returned home after the 1989 Soviet withdrawal, bin Laden was regarded as an Islamic hero. The 1990–91 Gulf War soon brought a new cause to his life when the royal family welcomed American troops to ward off the attack by Iraqi forces massed in Kuwait. Ironically, the Saudi government became a prime target of the fundamentalists the government had all but created. Like Dokubo Asari in Nigeria, bin Laden was outraged at the corruption of political and moral values that oil had wrought, and he stridently opposed the presence of an infidel army in

Islam's holy land. Dissatisfied and restless, he moved to Sudan and presided over a grab bag of small businesses started with his funds.

The Sudanese all but fleeced him, so he left for Afghanistan in 1996, nearly broke. Because he had spoken out against the Saudi alliance with America, family money was now out of reach. At times, bin Laden and his followers had nothing to eat in Afghanistan but stale bread. Al-Qaeda, mostly inert since its creation in 1988, was nursed to life with modest donations bin Laden raised on the basis of his previous exploits fighting the Soviets. His new enemy was the United States. Oil was now relevant not as a source of funding for bin Laden's activities but as the reason the United States kept its troops in Saudi Arabia (to ward off another invasion of Kuwait by the still-troublesome Iraq). Bin Laden's mission broadened beyond removing infidel soldiers from his homeland: he now wanted to cleanse the entire Muslim world of American and Jewish influences.

Kanan Makiya, an Iraqi intellectual, wrote a book about the terrors of Saddam Hussein's rule entitled *Republic of Fear.* The Middle East has many such countries, each different from the other, and Saudi Arabia is one of them. A hallmark of these countries, whether they are rich or poor, is a brutal security apparatus that, rather than elections or dialogue, is the method through which the unloved regimes deal with their opponents. Ibrahim al-Mugaiteeb, when I talked with him, was trying to avoid becoming the next victim.

Mugaiteeb had one of the loneliest jobs in Saudi Arabia. He was the oft-interrogated and occasionally imprisoned leader of the Saudi branch of Human Rights First, which consisted of a handful of activists sending out information and appeals to a world that paid little heed. As Amnesty International noted, "The Saudi Arabian government spares no effort to keep its appalling human rights record a secret, and other governments have shown themselves more than willing to help maintain the secrecy."

After arriving in Riyadh, I reached Mugaiteeb on his cell phone in Dhahran. He was parking his car at home and told me he had been summoned for a police interrogation the next day.

"I have been under scrutiny for the past two years," he said, his voice a mixture of impatience and nervousness. "All they have done is harass me. I do not know what they want to accomplish."

Mugaiteeb had arranged to meet a small group of diplomats the next day, so the summons appeared to be a form of punishment and a means of stopping the event from happening.

Might he be jailed when he went in for questioning?

"It would add another honorable individual to the long list of activists who are in jail," Mugaiteeb replied. "We are not sheep. Believe me, I don't care. They"—he was referring to the regime—"like to break people. They think they are getting on my nerves, but they can go to hell. Look at the price of oil, and still people do not have what they need. They think they can steal the resources of the country and the people will keep quiet."

He spoke fast, like a man who knew the connection might be cut at any moment. He quickly climbed the stairs to his apartment, and I could hear him gasping for air. His heart was not good, he said. Mugaiteeb was in his fifties and had ulcers, sciatica and thyroid problems. His financial health was not robust, either. He was $150,000 in debt and paid for his cell phone and Internet connection with credit cards that were nearly maxed out.

Opening the door to his apartment, he was met by his eighteen-month-old daughter. *Sweetheart*, he said in Arabic, *sweetheart*. I could hear the daughter's playful voice, and I could hear him start to cry. I asked whether we should talk later.

"She is going to miss me if she grows up without her dad," he replied. He began to sob again. The next day he had to submit himself to a system whose tortures and deprivations he had devoted his life to uncovering. He wasn't sure he would even have that long, because the police could come at any moment if they realized he was talking to a reporter. He spoke as though to a confidant, even though we had never met.

"We have to make our lives meaningful. People die by the thousands every day without doing anything. We must make meaning."

He paused.

"I am sorry for talking so much. I am very tired. I was beaten up two months ago."

He said his wife had just arrived home.

"She does not know the news yet," he whispered.

The youth of Saudi Arabia are children of oil. In 1973, the population was six and a half million, but the infusion of postembargo affluence helped fuel a demographic explosion. Baby boom followed baby boom, and the population is now about four times larger than it was thirty-five years ago. These Saudis would seem to have it made.

The latest boom was well under way when I visited the country in 2005. Riyadh's skyline was, at that moment, a panorama of cranes and extraordinary new buildings, including the Kingdom Center, a thousand feet high and shaped, at its top, like a postmodern bottle opener. Nearby, the Al Faisaliah Center had been designed in the form of an elongated pyramid, with a sphere housing a restaurant at its apex. The government was planning to build entirely new cities and further alter Riyadh's skyline with the $480 million Al Rajhi Tower, in the shape of a sail. The architectural brashness, mirrored in other countries and emirates of the then-booming Persian Gulf, made the vistas of Chicago, San Francisco and even New York seem tired and dull. And what was inside these buildings was remarkable, too. I visited the lobby of Al Faisaliah one morning and chatted with a real estate agent who was selling million-dollar vacation villas on man-made islands off Dubai whose collective shape resembled a palm tree. Sales were brisk as well-to-do Saudis, dressed in impeccable white *thawbs*, inspected a scale model of the fantasy islands. Actually, the islands were real—they were under construction, and the first ones would be completed a year later—but the most striking architectural statement, when oil prices fell as the global economic crisis took hold in 2008, was the forest of immobile cranes throughout the Persian Gulf. The construction of skyscrapers and cities and fantasies had stopped cold.

The illusion of everlasting wealth was carried along (yet undermined) by the fact that foreigners held two-thirds of all Saudi jobs (and an even higher percentage of private-sector jobs). Bangladeshis and Sri

Lankans labored at construction sites, Egyptians and Sudanese populated service industries, Americans and Britons ran the oil industry and banks. It was possible to spend months in Riyadh without speaking to a Saudi. Drivers, waiters, shopkeepers, bankers, nurses and doctors were almost all foreigners. The man who was selling the villas in the lobby was not Saudi. This would seem a benefit of possessing the world's largest oil reserves, because an army of foreigners can be paid to do your work. Saudi Arabia was one of the last nations on earth to outlaw slavery, in 1962, so it was accustomed to a subclass of workers.

But in fact Saudi Arabia, even in its boom days, had a severe economic problem. More than 30 percent of Saudi men were jobless, and they were not pleased, because few had a prince's stipend. Of course, this seems odd. Why didn't they just take the jobs that foreigners had? The answer was simple: most Saudis would rather be unemployed than accept low-paying, difficult work. And the good jobs—the ones in banking and other well-paying sectors—were out of reach for most Saudis because their education was insufficient. The elite sent their sons and daughters abroad for university and secondary school because the kingdom's recently formed educational system—its oldest colleges date from the 1950s—produced third-rate graduates whose studies were controlled, until a few years ago, by religious leaders who continue to exert a significant amount of influence. Even in the region, Saudi graduates could not compete: Egypt produced better engineers; Lebanon churned out better bankers. The best Saudi geologists did not graduate from King Fahd University of Petroleum & Minerals but from the Massachusetts Institute of Technology and the University of Texas. The government tried to force businesses to employ locally educated youths, but the program hasn't gone well. An American who ran a midsized engineering firm in Dhahran told me of a Saudi he reluctantly hired for a white-collar job who had a habit of assigning his work to expatriate Asians in the office. Making the situation worse, the Saudi did not listen to advice or instructions from the better-skilled Asians. He was not just worthless but disruptive.

These sorts of problems help explain why the country didn't have thriving industrial or technology sectors. The government had no shortage of funds to invest in new industries but lacked an ingredient that successful economies like China, Germany and Chile tend to possess: educated and motivated workers. And, unfortunately, the industry Saudi Arabia did possess—oil extraction and refining—was a miserly one when it came to job creation. Outside Dhahran I visited the corporate and residential compound of Aramco, the state-owned firm where the pay and benefits were wonderful but only for the lucky few with jobs there—and many of them were foreigners.

The compound was located at the spot where, in the 1930s, Max Steineke and his team of geologists had pitched their tents to explore for oil in the empty desert. Several million people now lived in the region, and the "camp," as it was still called, had the feel of a gated community. Its Rolling Hills Country Club had a golf course and a clubhouse serving burgers and salads. There was a bowling alley, duck ponds, baseball diamonds, parks with gazebos and grills, a horse stable called Hobby Farm, a skateboard park, a movie theater, tennis courts, a miniature golf course and miles of jogging paths. The streets had bucolic names like Prairie View and Eastern Avenue, and they were plied by yellow school buses and carefully driven SUVs that made full and complete stops. The ranch-style homes whispered "America" into your ear, as did the lawns and the kids on bicycles.

The white-collar residents of this enclave, who were a mixture of Saudis and expatriate Americans and Europeans, called themselves "Aramcons." Their lives were apart from what lay beyond the gates. In the rest of Saudi Arabia, Islamic custom prevailed, overseen by the *muttawa*. Women in shorts did not jog on the streets in the rest of the country, where they wore body-and-head-covering abayas when not at home and were forbidden from socializing with men who were not their fathers, brothers or husbands. Movie theaters were banned outside Aramco's gates, because reproducing the human image was still frowned upon (with the politically useful exception of the country's monarchs, whose faces and names were everywhere). The camp had a high-tech hospital that was far better than all but a few medical centers in the entire country. (Even so, when first-rate treatment was required, a private suite at the Mayo Clinic would be reserved for Prince X or Princess Y.)

My host was a Saudi employee of Aramco who was proud of the comfortable and enlightened life the company had created for its workers. After lunch in a camp restaurant that had the feel of an Ivy League faculty club, and that was staffed by Filipinos and Sri Lankans, we got into his car and drove toward Dhahran, a few miles away. There were no more lawns or even sidewalks along the roads, no women chatting with men or driving cars (the camp is one of the few places where women can drive in Saudi Arabia), no more of the quiet contentment that was the shared condition of Aramcons enjoying their well-paid jobs, excellent health care and competitive schools that prepared their children for modernity rather than joblessness or jihad. As I have said earlier, it is one of the oil industry's structural tragedies that it requires few workers, so there was little room for Saudis at their own oil company. In a country of more than 20 million people, Aramco employed 50,000, and each year it hired just 500 new employees.

"We are rich, but rich in what?" my host asked, downcast, as he dropped me off. "It is oil. It is not factories."

But there isn't even enough oil.

It sounds ironic or impossible, but Saudi Arabia, which has far more oil than any other nation and tends to be regarded as an expanse of wall-to-wall multimillionaires, does not have enough oil to make everyone rich, even when prices are high.

The crucial thing is not how many barrels of oil a country sells every day, but how many it sells *per capita*. For example, Kuwait's exports are one-fifth Saudi Arabia's but Kuwait's population is ten times smaller. On a daily basis, Kuwait produces more than a barrel a day of oil per person, whereas Saudi Arabia produces just about half a barrel. By this measure, Saudi Arabia is not even among the top five petrostates. The monarchies of Kuwait and Brunei can squander as much money as humanly possible—weddings can include almonds that are coated in gold flakes—and still there will be enough money for their nonroyal compatriots to live an easy life. In Saudi Arabia, oil revenues are sufficient to enable the royal family to build vast palaces for themselves and impressive highways for the masses, and to tax no one, and to offer education and health care subsidies, but insufficient to underwrite a consistent level of high comfort for all. In the early 1980s, per capita income reached $28,000, which ranked Saudis quite high in the world at the time, but it soon collapsed, along with oil prices, to one-quarter of what it had been—one of the most precipitous drops in national income in the twentieth century. It shot up again in the 2008 boom that pushed oil to nearly $150 a barrel, but then it dropped again as oil prices did. It was a roller coaster.

In a way, the grandness of the country's infrastructure was akin to the boast of a con man. King Fahd International Airport encompassed some three hundred square miles (though most of it was unused desert), had its own desalination plant, a mosque for two thousand worshippers and greenhouses. It was the largest airport in the world by surface area and, in fact, was larger than the *country* of Bahrain. Yet KFIA's main terminal was probably the emptiest airport in the world. A Saudi friend who met me there one day pointed out its wasted opulence as we walked through a vast air-conditioned garage that was all but devoid of cars. His laughter echoed off the walls, which had been built, like the rest of the multibillion-dollar airport, by Bechtel. These edifices reminded me of a complaint I'd heard from a dissatisfied prince in Riyadh who'd pushed for economic reforms that had not been adopted. "They are not nation building," he'd said of his own family. "They are building buildings." And when oil prices dropped in 2008, even the building of buildings came to a near halt.

Had oil done much good for the country? It might have been a strange question, but I tried it out on Bassim Alim, a lawyer in Jeddah who'd signed a controversial petition for greater openness in the kingdom and who was defending several colleagues thrown into jail for demanding more democracy. Alim wasn't a critic of the liberal, whiskey-sipping variety. Educated at Harvard, he wanted Islam at the center of Saudi life. In his view, that didn't mean the country had to deprive itself of political or economic pluralism.

We met at Casper & Gambini's, a café with the feel of a Manhattan restaurant, minus the alcohol. The decor was urban chic, with silver air ducts, polished wood, exposed brick and electronic music that emitted a sophisticated vibe. The place was filled with stylish young men and women. Because this was Saudi Arabia, where unmarried men and women are forbidden from mixing, the men sat downstairs and the women were upstairs, where there was a separate entrance. They could not see each other, but text messages of flirtation whizzed between them; occasionally there would be a round of laughs at a nearby table as one of the youths passed around his cell phone to show the latest missive he had sent or received.

"Oil has done nothing for this country that we couldn't have done without oil," Alim said. "I don't believe that in this region there is a nation that received all this wealth and squandered and looted it the way it has been done in Saudi Arabia. Cairo has an underground transport system and fourteen different universities. Cairo is poor, it is struggling, but it has a well-developed media and academic structure. There are political parties, though Mubarak is an asshole. There are protests, and society is vibrant."

Steak sandwiches were brought to our table by a Lebanese waiter. (Saudis would not take such jobs.) I was thinking to myself that if Egypt was enviable, something was badly amiss. Alim explained that his jailed colleagues had not gotten into trouble for signing a give-us-the-head-of-the-king document; their petition merely asked for greater popular participation in the affairs of government. The king had a Consultative Assembly, the Majlis, whose members were appointed and infrequently summoned. Token municipal elections had been held, but the winners had almost no power. The king decided how much money would be allotted to each ministry, and outsiders were not allowed to look into allegations of waste or abuse. "No one can dare ask the government for accountability," Alim said. "Where is this money going?"

We finished our meal at around eleven in the evening, when the night was still young. The rhythm of life in Saudi Arabia was as particular as its demographics: 75 percent of the population was under the age of thirty. Because so many young men didn't have jobs or put in just a few hours at work that meant little to them, and because the heat discouraged daytime socializing, the night was when Saudi Arabia came alive. As Alim and I left Casper & Gambini's, the young men around us, dressed in jeans and crisply ironed shirts, continued their texting activities.

These men's disassociation hardly took the form of jihad; if they died young, it would be from another form of self-harm. Addled on illicit drugs or simply bored, young Saudis were among the world's most dangerous drivers; more than 80 percent of the deaths in state hospitals were due to road accidents. All of a sudden, without its blinker flashing, a car would swerve in front of me or overtake me on the right-hand side, using the shoulder of the road. Youths, indulging in a practice known as "drifting," gathered on lightly used roads and spun their cars into sideways skids at high speeds, sometimes with fatal outcomes. I was not surprised to see a YouTube video showing a car on a Saudi highway with its doors open and several youths holding on to them, with nothing more than sandals on their feet, surfing along the pavement.

I was more interested in finding a youth who could reveal the other, more violent side of alienation. Mohammed Ibrahim Abdul Aziz fit the bill.

When our paths crossed, Mohammed was twenty years old. A native of Dammam, in the country's oil region, he seemed young for his age— sparse and silky facial hair gave him the look of a teenager. He had studied at King Saud University, an impressive institution from the outside only. The new campus, built in the 1980s, cost about $4 billion (adjusted for inflation, more than $8 billion today), but its Saudi and foreign professors were unknowns in the academic world. Mohammed, neither inspired by his teachers nor hopeful of getting a job if he completed his studies, dropped out and joined the generation of highway-surfing youths.

Like many young Saudis, Mohammed spent his spare time cruising the Internet, which, though censored by government filters, served as an outlet to most things illicit—from online chats with girls to, of course, jihad. When I asked which fundamentalist Web sites he visited, Mohammed couldn't remember precisely, because there were so many, all easily available, extolling the glory of fighting for Islam in, for instance, the Palestinian territories against the Israelis and in Iraq against the American occupiers. It was the latter fight that drew Mohammed's interest, after a fundamentalist friend suggested that he devote his life to the intensifying war against the infidels there.

I met Mohammed in Iraq in the spring of 2005. He had been captured a few days earlier by Iraqi security forces. As it turned out, Mohammed was a happy prisoner because his time in the insurgency had been unhappy; the jihad was not, he realized, all it was cracked up to be. Because he was captured about seventy-five miles north of Baghdad in Samarra, where I happened to be reporting a story, I was offered an opportunity to interview him. In the end, Mohammed provided a better glimpse into the mind-set and future of his people and his homeland than the oft-quoted and always-followed oil minister I'd encountered in Washington. In sketching his life and his journey, Mohammed drew a portrait of how oil paid for and provided the inspiration for his violence.

Mohammed had seen better days. He was wearing a green shalwar kameez covered in mud, and his eyes were bloodshot. He had been interrogated almost nonstop by Iraqi and American soldiers and had willingly provided answers, according to Mohammed himself as well as his interrogators. A soiled bandage was wrapped around his head; he said he'd been injured when the car he was traveling in, with two members of his insurgent cell, was attacked by Iraqi soldiers. It was just as probable that he had been roughed up but did not want to say so. We talked in a small office in Samarra's ad hoc detention center; a desk in our midst had bloodstains down its side. In normal times, the detention center was Samarra's library, and an Islamic inscription over the entrance announced, "In the name of Allah the most gracious and merciful, Oh Lord, please fill me with knowledge." Throughout my talk with Mohammed, audible from parts of the detention center I was not allowed to visit, prisoners screamed and retched.

The Saudi regime no longer abets its youths' taking up arms against infidels, but Mohammed's passage to Iraq was not so difficult. It began, he told me, with reading and listening on the Internet to sermons by fundamentalist imams. Many of Mohammed's friends supported jihad, and one of them offered to put him in touch with someone who could facilitate a journey to Iraq. After a friend was killed in a car accident, Mohammed began thinking about the aimlessness of life in Saudi Arabia. He was like dry timber to a match; his life would gain meaning, he decided, through a martyr's death in Iraq (though not by suicide bombing).

The facilitator gave Mohammed the phone number of a man in Damascus who would spirit him into Iraq. Mohammed memorized the phone number. Once in Damascus, he was instructed to take a bus to the border (he bought his own ticket). There, he met up with a smuggler who demanded $50 and took him into Iraq. After they crossed the border, local insurgents took his passport and the rest of his money, about $250.

Mohammed's career as a holy warrior lasted a few weeks. He had no skills to offer the insurgency, because he had never fired a weapon or built a bomb, did not know his way around Iraq and could not even blend into a crowd, because his Saudi accent gave him away. When he realized that his insurgency cell was led by a man who seemed more interested in stealing cars than killing Americans, he wanted out. His capture came as a relief, which is why he had not been tortured to the edge of death—he was more than happy to tell everything he knew.

"I made a mistake," Mohammed said. "I just hope I will be allowed to go back to Riyadh. I want to leave."

He would not be going home soon. An American military adviser dressed in jeans, a former special forces officer, was monitoring my talk with Mohammed. The American had a pistol strapped to his thigh. The interpreter, with a pistol on his hip, was an overweight Iraqi police official. The Saudi, the American and the Iraqi in this room were in a deep mess, as were their homelands. There were many reasons, and the core one was evoked when Mohammed ventured a guess as to why Iraq had been invaded.

"The Americans want to control Iraq's resources," he said. "They came here for oil."

9

Empire

One day, Vladimir Putin visits a gas station in Chelsea.

No, this is not a setup for a joke.

The station is at the corner of Tenth Avenue and Twenty-fourth Street in Manhattan, where it draws a steady flow of yellow taxis. On September 26, 2003, this intersection attracted instead a fleet of black SUVs with tinted windows, out of which emerged Secret Service agents in dark suits whose semicovert earpieces suggested an event of unusual significance. A ribbon-cutting ceremony was being held at the just-renovated station, and the VIPs included Senator Charles Schumer and Vagit Alekperov, the Russian billionaire whose company, Lukoil, had bought Getty Petroleum's chain of thirteen hundred gas stations. The pumps at this station were new, as were the paint and signs and flags, bearing the name and red-and-white colors of Lukoil.

The howling of police sirens heralded Putin's approach. When he emerged from his limousine Schumer and Alekperov attached them-selves to his side and guided him to the station's mini-market, where the president ate a Krispy Kreme donut that had a caloric value at odds with his semifamous judo fitness routine. Approaching a pod of TV cameras, Senator Schumer took the initiative, living proof of the insider joke that nothing is more risky than standing between Chuck Schumer and a microphone. With gasoline prices rising at the time, and with the Middle East becoming a geographic profanity in America, Schumer wanted to thank his new best friend.

Vagit Alekperov, president of Lukoil

"We are welcoming Lukoil to New York because we want to see competition," he said. "The more competition there is, particularly against OPEC, the better New York will do and the better America will do."

Schumer beamed broadly and even bowed slightly as he pumped Putin's hand. The latter, dressed in a statesmanlike gray suit, white shirt and blue tie, smiled without warmth, which is harder than it sounds, and offered his hand in a programmed way. Known for being abrupt, Putin departed without making a public statement. His presence was his message. Alekperov articulated a scrubbed version of the view from the Kremlin. "American consumers are getting a new source of energy," he declared. "The source is located on reliable Russian territory."

Russia's resurrection at the dawn of the twenty-first century is one of the great narratives of our times. It goes like this: After the collapse of the Soviet Union in the early 1990s, Russia was not so much a country as a black-market bazaar stretching across eleven time zones in which anything could be purchased—a favor from a minister, a murder by a cop, a kilo of uranium. Boris Yeltsin was a brave and well-meaning president, yet a drunk in charge of a collapsed economy. Then, the miracle. Oil prices began to rise and within a few years the nearly 10 million barrels a day that Russia pumped out—it was the second-largest producer in the world, after Saudi Arabia—netted a new fortune. Instead of spiraling into the status of a nuclear-armed failed state, Russia regained its health under Yeltsin's handpicked successor, Vladimir Putin. That's why Putin, rather than dashing around New York to beg for aid, was instead celebrating a Russian company's purchase of a slice of corporate America.

What's accurate about this narrative is that oil had pumped Russia full of money, turning it into an energy superpower. By 2004, Moscow counted more billionaires than New York, and middle-class Russians who could not afford a new pair of shoes in the 1990s were flying to Greece on package vacations. But where was Russia heading? The local media, remarkably independent in the Yeltsin era, were all but nationalized by Putin. The opposition, ignored or maligned on the airwaves and harassed in the streets, became irrelevant. As democracy shrank, corruption thrived, life expectancy continued to decline and the security services occupied the commanding heights of not just law enforcement but economic activity. After the inevitable easing of its oil boom, what kind of empire will Russia finally resemble, if it resembles an empire at all?

I had a particular interest in this question. In college, I studied Russian and enrolled in as many courses in Soviet and Russian history as I could find. I even attended a summer program at Leningrad State University, which was Putin's alma mater, though he had entered the KGB by the time I arrived in his hometown in 1980. What was missing in my studies was oil. I started to fill the gap once by chance in my senior year, when I visited the Rialto theater in Berkeley, California, to watch Andrei Konchalovsky's *Siberiade*, an epic about Siberia and its alteration after oil was discovered. I enjoyed the film, but understanding the Soviet Union and the Cold War did not seem to require a grasp of oil's role. Brent Scowcroft, national security adviser to President George H. W. Bush and an eminence of the Cold War establishment, manifested this blind spot when I asked him a few years ago about oil in Soviet history. "Let's suppose the Soviet Union was like India, without huge energy resources," he mused. "It might have made a huge difference. I never thought about it."

The few people who have thought about it rightly concluded that oil was crucial to the life and death of the Soviet Union, and their findings are relevant to what is unfolding in Russia today. In the late 1950s, the Soviet economy faced an energy crunch; it even had to import some fuel. Then oil was found in Siberia. Moscow could not only fill its own energy needs, it could give oil to its Communist allies and earn cash by selling some of its crude on open markets. When the 1973 embargo initiated high oil prices, Soviet output was fortuitously hitting its stride, eventually reaching 12 million barrels a day, which was more than Saudi Arabia produced. It was a regime-extending windfall for the Kremlin. I remember how Americans always marveled at the inferiority of Soviet products and the mystery of how the USSR kept its tottering command economy going. But it wasn't so mysterious—80 percent of the Soviet Union's considerable hard-currency earnings came from energy sales, just as most of Russia's export revenues now come from the same source. In Soviet times, this financial adrenaline sustained the state but created a vulnerability, because if the oil money ceased to flow, the dysfunctional corpus that was the Soviet economy would crash. "Without the discovery of Siberian oil, the Soviet Union might have collapsed decades earlier," wrote Stephen Kotkin, director of Russian and Eurasian studies at Princeton University.

The Central Intelligence Agency realized in the 1980s that nuclear missiles were not the only doomsday weapon that could destroy the Soviet Union. A steep fall in oil prices might do the job without radioactive consequences. When Saudi Arabia's King Fahd visited Washington in 1985, the price of a barrel was $30, a very high level (in those days) that was painful for consumers. President Reagan urged Fahd to lower oil prices, which could be done through higher production. Reagan's stance was shaped largely by domestic concerns—Americans would have more money in their pockets if oil cost less. But the impact on Soviet finances was emphasized by CIA director

William Casey, who met secretly with the Saudi ruler to promote the virtues of low oil prices as an economic dagger into the hearts of the godless Communists in Moscow.

As it happened, King Fahd was upset with his OPEC comrades, all of whom sold more oil than allotted under their quotas. It was time to rap them on their knuckles. At the end of 1985, the Saudi monarch raised output dramatically and by the middle of 1986 a barrel cost just $12. This occurred as Mikhail Gorbachev came to power and tried to reform the dying Soviet system. "Gorbachev's incipient perestroika was instantly fleeced of much of its hard-currency receipts," Kotkin noted. According to Yegor Gaidar, who oversaw the economic "shock therapy" that accelerated Russia's transition to capitalism during the Yeltsin era, the fall in oil prices deprived the Soviet Union of $20 billion a year—"money without which the country simply could not survive." As Casey certainly celebrated at CIA headquarters, cheap oil doomed the Soviet regime. A political generation later, it's possible history will repeat itself.

Six months after Putin's Krispy Kreme moment, I visited the Moscow headquarters of the company that owned the presidentially inaugurated gas station in Chelsea.

The tower that housed Lukoil were built to intimidate rather than please. It was impossible to see inside the glass windows, tinted in a menacing shade of zinc and grime, and the contours of the structure discouraged long-lasting gazes—the tower was a brutal dagger that jutted into the sky. But inside, on the executive floor, the offices were as expensively outfitted as any of the citadels in Houston. The desks were made of tasteful hardwoods, the track lighting cast just-so amounts of light and the Prada suits worn not just by executive vice presidents but also by their secretaries indicated a pleasure and pride in letting visitors know that sartorial choices at Lukoil were not affected by price tags.

Vagit Alekperov, the president and largest single shareholder, whom I mentioned in chapter 6, had evidently gone to the same anti-charm school as Putin. Alekperov's laser stare could melt a glacier. His hair was a few millimeters longer than a crew cut, and with his stocky frame, he had the hard look of a high-priced thug. He spoke in a voice that often slid into a mumble, as if it was not his job to speak clearly but yours to listen closely. Most people chose to listen closely, because his company presided over more oil than Exxon and employed 150,000 people. Alekperov's personal fortune was in the billions of dollars, so subordinates scurried in a manner that evoked a royal and fear-inspired court.

The oil world is known for rough characters—Texas wildcatters are as tough as the prairie and proud of it—but Alekperov made his American cohorts look like sissies. At the start of his career, in the 1970s, Alekperov lived on primitive offshore rigs in the Caspian Sea that routinely suffered blowouts and erupted into flames. An explosion once threw him into stormy waters; he had to swim for his life. In the late 1970s he was sent to Siberia, where, the story goes, a pipeline ruptured and repairmen refused to get near it, fearing an explosion. Alekperov sat on the pipe so his reluctant workers would gain some confidence. His do-what-it-takes attitude also applies to ethics. When negotiations stalled over access to Kazakhstan's Tengiz field, Alekperov gave the Kazakh president a $19 million executive jet. Asked by a reporter whether any strings were attached, Alekperov replied, "Nothing is free." Lukoil won a chunk of the Kazakh project.

Because crude oil and political power are umbilically connected in Russia, Alekperov was nonetheless answerable to a higher authority. I learned this within moments of entering his office.

With an expansive view of Moscow, Alekperov's quarters were furnished in a severe, Germano-Scandinavian fashion, the floors and walls covered in a polished cut of blond wood, and the couches and chairs upholstered in red leather. A large relief map of the world hung on one wall, with small lights marking the places where Lukoil had investments. (The map had many lights.) On the wall behind Alekperov's titanium-and-glass desk, on which there was not a stray paper, hung an ornate carving of Russia's coat of arms, which features a double-headed eagle. Owing to the map and the official seal, the office felt like the work space of a general or a politician, and this impression was confirmed by the sole picture on Alekperov's desk. The color photograph did not show his wife or teenage son. It showed Vladimir Putin.

If the coat of arms had been replaced with a hammer and sickle, and the photo of Putin swapped with Brezhnev's, I could have imagined myself back in the USSR, and so could Alekperov, who was the Soviet Union's last energy minister. Now, instead of laboring on behalf of Marxism-Leninism, Alekperov worked for Russia, Inc. The name was different, but the new institution functioned in a similar way to the old one. Alekperov understood that in Russia, as in the Soviet Union, the Kremlin was the center of economic as well as political power. The Kremlin's wishes were to be obeyed, not questioned.

"Politics are close to me, but there are different ways of participating in politics," he told me, speaking carefully. "I don't have personal ambitions. I have only one task connected with politics: to help the country and the company." He nodded affirmatively when asked whether he regularly met Putin. He said, in a tone of unusual reverence, "I treat him with great respect."

On Tverskaya, the historic boulevard that spills into Red Square, I waded through money. Designer boutiques lined the street and were tucked into the lobbies of five-star hotels that charged nearly any price they desired. The road was a conga line of late-model BMWs and Mercedes-Benzes, some with bulletproof doors and windows that were necessary precautions against assassination. Murders for hire still happened all the time, due to lethal rivalries for control of Russia's wealth that had begun in the 1990s. Draped in the must-have accessories of the nouveaux riches—Gucci, Hermès and Tiffany—the men and women who descended from these gleaming and armored sedans could have been on Rodeo Drive for all the cosmetic surgery and distance from ordinary life they projected. There seemed to be sushi bars every few yards, in a city that just a few years earlier, during my last visit, had had the dismal ambience of a soup kitchen running low on soup. Now it was so different, thanks to hundreds of billions of dollars in oil revenues flowing into the wallets of an elite that had already run out of sensible ways to spend its wealth. America's gilded age had nothing on this. Roman Abramovich, an oil multibillionaire, acquired for personal use not only a $100 million Boeing 767 and several yachts costing more than $100 million apiece, but a British soccer club that passed into his hands for $230 million. He explained his investment strategy as "I love this sport. . . . Why don't I get my own team?"

I entered the glittering lobby of the Marriott Hotel on Tverskaya to meet one of the few people sounding an alarm. Andrei Illarionov was an improbable candidate to shout out that the emperor had no clothes (or too much oil), because he was Putin's senior economic adviser. Married to an American, Illarionov was a passionate fan of Ayn Rand, who wrote *Atlas Shrugged* and other celebrations of the blessings of free markets and the evils of government regulation. (Illarionov had presented Putin with a collection of Rand's works.) Though Moscow did not lack for laissez-faire economists who treasured Thatcher as much as Pushkin, Illarionov was the best and the drollest. His jowls and thick voice gave him the look and sound of Rodney Dangerfield with an economics PhD. He was an anomaly in a regime led by a former KGB agent whose doctoral thesis was titled "The Strategic Planning of Regional Resources Under the Formation of Market Relations."

We found an out-of-the-way couch in the lobby and got down to business: a soliloquy by Illarionov about the ruination of Russia's future. The problem consisted of the way oil and gas revenues were being used by Putin to enlarge his personal power and that of the government. Illarionov noted that the police had recently arrested Mikhail Khodorkovsky, an oil billionaire who backed the opposition; Khodorkovsky's company, Yukos, was being seized by the government. Illarionov, who had begun to openly criticize the regime in which he served, described the Yukos takeover as "the swindle of the year." He accused Putin of creating a "corporate state" that stifled individual initiative and the growth of a diversified private sector. When the injections of petroadrenaline plateaued, as they would one day, Russia would have little else to keep its economy going, Illarionov argued.

The president's economic adviser was, it turned out, a dissenter.

Illarionov was late for a dinner at Prague, an ornate restaurant a few miles away, so we jumped into his government-issued Audi and continued talking as his driver weaved through the gridlock. Thanks to its new affluence, which put vast numbers of cars on the roads, Moscow at rush hour was far worse than Manhattan. Despite Putin's law-and-order reputation, the streets were like Chicago's of the Capone era. A few days earlier, a business executive had been killed near the Marriott, along with his driver and bodyguard, when a motorcyclist put a bomb on the roof of his armored Volvo. What was remarkable about this assassination was that in the previous year this executive had survived an assassination attempt on the very same street. Where the Mafia's violence didn't reach, the police state's arm was felt. I dined at a posh restaurant one evening with a banker who needed, midcourse, to make a business call. He faced a problem because he knew his cell phone was bugged and he assumed mine was, too. So he borrowed a phone from another diner, a total stranger. The diner offered his phone without a blink; he no doubt asked the same favor of strangers, too. (It would not matter if the stranger's phone was bugged, because the eavesdroppers would have no idea who was making the call.) It was a merry-go-round of precaution, and for the city's monied elite, this was completely normal. Prudence is not unusual in business circles, but these gestures stemmed from fear rather than competition.

By the time we arrived at Prague, Illarionov was arguing that oil allowed Putin and other strongmen to claim responsibility for a tide of prosperity that only coincided with their rule. This amounted to a lesser version of the resource curse, because Putin and Hugo Chávez were not enemies of their people in the fashion of Equatorial Guinea's Teodoro Obiang or Angola's José Eduardo dos Santos, who for motives of greed, fear or general inhumanity were hoarding their nations' wealth and killing or imprisoning those who disagreed. Critics could say what they wanted of Chávez, but he cared about Venezuela and was not torturing opposition leaders. Putin reportedly amassed a personal fortune while in office but was a nationalist who wanted Russia to be strong. The non-cataclysmic condition of their nations was also due to the existence of middle classes that acted as brakes on the worst excesses of dictatorship.

This does not mean Putin and Chávez were making the right decisions. Illarionov's attention was fixed on the Aladdin-like powers that oil imparted to these men. "There is a paradox," he said as we walked into the restaurant, where a banquet had begun for a group of Western economists to whom Illarionov was a rock star, or as close as the world of Hayek and Friedman gets to heart-throbbing adulation. "What do you use this power for? Do you use it to reduce the size of the government, to break monopolies, to make the country more competitive and liberal? Or do you use this power to just preserve your own power?"

Since Putin had come to office, the state had acquired control of NTV, an influential, privately owned television station that broadcast nationally. Most of the independent newspapers had been acquired by firms loyal to the state. Russia was becoming the third-most-dangerous country for independent journalists, ranking behind only Iraq and Colombia in the number of murdered reporters. Opinion polls showed Putin's popularity as quite high, but it would be hard to be unpopular when the media laud you every day and oil prices as well as your production of oil are rising fast. It did not matter whether the emperor had any clothes—whether, in other words, the country's economy was becoming softer rather than stronger. What mattered, at this exuberant moment, was whether the emperor had money to dole out and a subservient press to take note of his largesse.

Illarionov posed his doubts at a time, in 2004, when it was possible to believe that Putin might still listen to smart advice that did not come from his colleagues in the security services. I wanted to ask Illarionov for a final prediction of the way things would turn out, but the doors to the banquet hall had just been opened. Illarionov was swallowed by the din of a celebration that was far from over.

"Power" is a vague word. It can mean gasoline or knowledge or the authority to order an execution. Oil is unique because it can be transformed into so many types of power—it can make lights burn and planes fly; it can turn wildcatters into millionaires and nations into superpowers. This is a problem as well as an attribute, because when power is measured in the hundreds of billions of barrels, and when it can be controlled by one man or one institution, it can be too much of a good thing.

In Russia, as elsewhere, if the state does not control oil, oil can control the state. It's a dilemma: What is better, for oil to be in the hands of companies that become immensely powerful or for the government to own everything and become the center of all things? Illarionov's free-market argument was sympathetic, but his Randian alternative—oil in the hands of the Russian private sector—was in this case a cause for concern. This was not a theoretical issue, because Putin was at that moment cracking down on Russia's wealthiest oil baron. In this story, there are no heroes or villains, just problems that few countries resolve.

At the outset of 2003 Mikhail Khodorkovsky was the largest shareholder of Yukos and the wealthiest oilman in Russia—even richer than Alekperov. Unlike Alekperov, who could not respond quickly enough to the Kremlin's wishes, Khodorkovsky had his own ideas. Khodorkovsky's motives were not entirely noble—he was upset by Putin's desire to collect more corporate taxes and reduce the clout of oligarchs like himself. After acquiring a chunk of Russia's oil reserves, Khodorkovsky was now using his wealth to fund a chunk of the opposition. But Putin was gutting the courts, the media and the opposition, so Khodorkovsky was also genuinely concerned about Russia's democracy. He had amassed a fortune in his early thirties by taking risks that went his way, so his challenge to Putin was just the latest gamble that he expected to turn out well.

Khodorkovsky was also challenging Putin in another way: he had opened talks with Exxon to sell a large slice of Yukos to the American firm. This was a provocation, because Putin did not want an American oil company to own the largest Russian oil company. Putin's concern was understandable. When a Chinese firm tried to buy Unocal, an American oil company, the outcry in Congress was immediate; the sale did not go through. So it wasn't a great surprise that when Khodorkovsky's private jet landed at a Siberian airstrip one day, paramilitary officers stormed aboard and arrested him. Hauled back to Moscow, Khodorkovsky was accused of tax evasion and fraud. In court appearances, he sat in a cage.

My visit to Moscow coincided with his trial, so I met his lawyer, Anton Drel, at one of the mansions that had not yet been seized from the jailed billionaire. As we sipped tea under stained-glass windows that seemed right for a cathedral rather than a home, Drel described the trial as an emblem of Russia's undemocratic course. Even Drel's office had been raided several times. His jailhouse consultations with Khodorkovsky were assumed to be bugged. Drel and his client silently exchanged notes that they scribbled over after reading. "Like in the movies," Drel said.

The case was hopeless, partly because Khodorkovsky, like any businessman of that era, deserved to be convicted—though for other crimes. Under the post-Communist rule of Boris Yeltsin, lawlessness pervaded business transactions as the government sought to break apart the Communist-era state-dominated economy. Instead of having transparent privatizations, state assets were stolen or stripped by managers and entrepreneurs who had the nerve or connections or assassins to get what they sought. The new generation of billionaires became more powerful than the state, whose belongings they carted away like a museum's unprotected valuables. Yegor Gaidar, an architect of this economic shock therapy, acknowledged when we talked in Moscow that the fire sale of state assets, particularly oil and gas holdings, created a class of oligarchs who ruled the economy and the state. "To some degree they regarded themselves as the real government of Russia, and to some degree they were the real government," Gaidar said. "They could easily dismiss ministers and nominate people who would be loyal to them."

The chaos of the early Yeltsin years had begun to abate by the time Putin came to power in 2000; the government was regaining its clout. Putin put the reaccumulation of state power into overdrive. Instead of independent businessmen running away with the gems of the economy, allies of the Kremlin did so. Putin's message to the Yeltsin-era oligarchs, transmitted in a variety of ways, including the prosecution of Khodorkovsky, was as sharp as a winter gust in St. Petersburg: even if you own every share and every paper clip of your company, you must follow my orders or lose everything. Vagit Alekperov had gotten the message: that is why, when I sat in his office, Putin's eyes also bore down on me.

Khodorkovsky is serving his jail sentence in Siberia, and the oil fields he once controlled are in the possession of a state-owned firm. The merging of oil power and state power was so thorough that in 2008, when Putin was obliged to hand over the office of the president after his second term—he sidestepped into the post of prime minister—he selected Dmitry Medvedev, whom several years earlier he had selected as the board chairman of Gazprom, the vast state-owned oil and gas firm.

The renaissance narrative pivots around the idea that oil and Putin were responsible for Russia's economic rise. Annual growth exceeded 6 percent during his presidency, and once-bankrupt Russia even set up sovereign wealth funds to save some of the oil bounty for future generations. But there is an alternative narrative, which is that Russia's economy should have performed even better than it did, that oil and Putin were a hindrance. And this alternative narrative was proven correct as Russia's economy, which seemed so fat and invincible when oil prices were skyrocketing, staggered under the blow of fallen oil prices in late 2008.

As Michael McFaul and Kathryn Stoner-Weiss, Russia experts, have noted, between 1999 and 2006 Russia's growth rate, impressive in isolation, was only ninth fastest among the fifteen post-Soviet states. When the Soviet Union fell apart, its successor states went into an economic free fall that halted after each government initiated emergency reforms, as all did. Russia, like every other country in the former Soviet Union, was experiencing economic growth before Putin came to power. "Putin arrived on the scene at a good time in Russia's economic cycle, and got even luckier as oil prices rose worldwide," McFaul wrote with Stoner-Weiss. They noted that Russia's standings in corruption and public health surveys worsened under Putin. Russia's fate rested once again upon a fickle resource sector that had functioned, in the Soviet era, as a trap door; once oil prices receded from nearly $150 a barrel, as they would, the country would find itself with a crippled economy and a brittle autocracy. "The strengthening of institutions of accountability—a real opposition party, genuinely independent media, a court system not beholden to Kremlin control—would have helped tame corruption and secure property rights and would thereby have encouraged more investment and growth," McFaul and Stoner-Weiss wrote. Not long afterward, when oil prices fell below $100 a barrel, the Russian stock market plummeted along with the value of the ruble, and the once-mighty government surplus, along with one of the sovereign wealth funds, were greatly depleted to support the faltering economy. Russia's megarich took a cold bath, with Moscow dropping below New York as the billionaire capital of the world. Even Putin's popularity began to shrink. A rash of street protests broke out—and were quashed.

As Russia suffers the blows of falling prices and declining output, McFaul and Stoner-Weiss worried that it could become another Angola, led by hard men who care more about controlling crude-oil money than providing good governance. An extreme outcome of that sort is not certain, thankfully, but McFaul, who in 2009 was appointed to President Obama's National Security Council, was not alone in seeing an unfortunate future. His conclusion was shared by Andrei Illarionov. A year after our encounter, which ended at the Prague restaurant, Illarionov turned on Putin and resigned. "The state has become, essentially, a corporate enterprise that the nominal owners, Russian citizens, no longer control," he wrote. "There are other countries like this: Libya and Venezuela, Angola and Chad, Iran and Saudi Arabia. Russia is one of them now. It is a historical dead end."

10

Mirage

The addictions of Hugo Chávez, president of Venezuela, are regularly in full view. On his television show, *Aló Presidente*, Chávez sips espresso from a white porcelain cup, and because the program can last from morning until night, with Chávez talking and singing and crying and joking and taking phone calls from Fidel Castro, the nation watches him drink cup after cup. Quite famously, the paratrooper-turned-president is wired on caffeine. That's not his only craving. In the halls of American power, Chávez is known as a leftist who appeared at the United Nations a day after President George W. Bush and proclaimed, crossing himself and sniffing the air, "The devil came here yesterday, and it still smells of sulfur." In his disobedience of political etiquette, Chávez acts with intended provocation. His defiance extends to the realm of economic strategies, because he is trying to overturn the dismal conventions of third-world resource management.

If, in the last century, you watched in dismay as oil profits were stolen or wasted, you might have been hopeful when Chávez was elected president and vowed to use resource wealth to help the needy. Though Venezuela has the world's seventh-largest reserves, most of its 26 million citizens are exceedingly poor. The enclaves of wealth in Caracas are surrounded by coils of angry slums. It is a classic example of what economist Joseph Stiglitz calls "rich countries with poor people." Chávez's desire for a fairer economic order was not new, because radical and well-meaning leaders across the globe had tried to make oil a blessing. Nigeria had had one or two presidents who preferred reform to looting, and even Huey Long tried to spread the oil wealth in Louisiana. But Louisiana remains one of the poorest states in America, and Nigeria is, well, Nigeria. I went to Venezuela to see whether Chávez could perform the magic that had eluded so many others, and my first stop was the barrio of Gramoven, where a new paradigm of resource management was being built.

Gramoven, at first glance, seems a model for little more than world-class squalor. Its crowded streets are lined with bare-essentials shops selling everything from sacks of flour to used shoelaces. Young men linger on corners in the way of the unemployed, swapping rumors about jobs that are hard to find. There is a wariness in their eyes, on the lookout for not just work but danger, because on these unkind streets even the jobless are mugged. Other hazards include manhole covers that have been stolen, which means that if you do not watch your step, you can disappear into a black hole. In a general sense, Gramoven is a black hole of poverty from which few escape.

A supporter with a placard of Hugo Chávez, president of Venezuela

Gramoven was hosting a vision of the future that went by the awkward name of Fabricio Ojeda Nucleus of Endogenous Development. The "nucleus," located on a side street near the barrio's heart, consisted of three main brick buildings the size of low-slung dance halls. One building housed a medical clinic, while the others held cooperatives that produced shoes and clothes. The well-tended complex covered just sixteen acres and had, at its center, a small amphitheater for meetings and performances; off to one side were an organic garden and a sports field. This nucleus was a model for Chávez's effort to plow oil money into social development. There were plans for hundreds like it across Venezuela, and not only did the funding come from oil, but the state-owned oil company managed everything. At the time I visited in 2005, the nucleus had received more than $7 million from Petróleos de Venezuela S.A., and a PDVSA manager, wearing a company badge, helped run the place. It was a showcase of sorts, because Chávez had broadcast an *Aló Presidente* episode from it, and its visitors included Harry Belafonte, Danny Glover and Cornel West.

The shoe cooperative, suffused with the aromas of leather and glue, was brightly lit and freshly painted. Its sewing and cutting machines were not crammed together, as they might be in a typical sweatshop. The pace of work was not hectic when I visited, and perhaps because of that, the output was a modest six hundred pairs of shoes a day. The measured rate of production did not translate into high quality, unfortunately. The workers were new to shoemaking and most of their output went to Cuba, which cannot afford to be picky, or was distributed at discount prices to poor families in the barrio. Oswaldo Quintero, one of the associates, as workers called themselves, explained that the 140 members of the cooperative voted on their pay (about $190 a month) and hours (two six-hour shifts a day). Quintero, who was forty years old, a former taxi driver and the father of five children, had an Everyman look, with a slight potbelly, a two-day stubble and short legs. His blue overalls were smeared in shoe polish. He savored his new life because he didn't need to drive around the city for twelve or fourteen hours a day, six days a week, risking robbery or carjacking every minute.

"PDVSA now belongs to all Venezuelans," he said. "Before it was just a small group who profited from it."

He meant that PDVSA, though state-owned, had not served the state well. In 1976, when a nationalization law went into effect, PDVSA gained control of the country's oil reserves. By the 1990s, most of the firm's gross revenues were plowed back into its operations; the rest went to the government. Because oil revenues were the government's largest source of income, the company ended up with a larger budget than the government, and this had the perverse effect of creating a prosperous first-world company in an impoverished third-world nation. The firm had a talented and well-paid cadre of engineers, its facilities had up-to-date equipment and it smoothly pumped out large amounts of oil—reaching a peak of nearly 3.5 million barrels a day. But like the foreign companies with which it operated joint ventures, PDVSA spent only a token amount of its considerable revenues on social or economic programs.

When Chávez was elected, PDVSA was quasi-independent of the government that owned it. This would not last. In 2002, after a series of political conflicts that included an anti-Chávez coup, PDVSA workers went on a two-month strike that ended with Chávez firing eighteen thousand managers and engineers—most of the firm's white-collar workers. Chávez proceeded to turn the firm's priorities upside down. Instead of about 40 percent of its revenues going to the state, two-thirds did. But there was a twist: instead of the oil money being transferred to the government and then to ministries that oversaw health, education and welfare programs, PDVSA was put in charge of the blitz of new programs. Chávez calculated that PDVSA's revamped staff would be more loyal and more capable than the civil servants whose uninspired presence lent government ministries the aura of early retirement homes for bureaucrats.

"Sowing the oil"—in Venezuela, this phrase is often used to describe the spending of oil revenues on human development—had a quick impact on the lives of people like Quintero. Thanks to his reasonable work hours at the nucleus, he had enough time to attend an adult-literacy course for which he received a "scholarship" of nearly $100 a month; he was being paid to take the course, which was held at a nearby school. PDVSA subsidized these courses—not only the scholarships but teachers, textbooks, televisions and videocassettes. PDVSA funded these adult schools across the country, as well as a network of new universities and secondary schools named after Chávez's nineteenth-century hero, Simón Bolívar, who fought for Latin American independence.

Because oil prices rose from almost the moment he was elected, Chávez was able to pour tens of billions of dollars into these programs. He had the same fortunate timing as another statesman who came to power as oil prices took off—Vladimir Putin. Chávez called his reform movement a Bolivarian revolution, and poor Venezuelans were its beneficiaries. For the first time in his life, Quintero even had access to decent medical care, thanks to the nucleus clinic, which had six pediatricians, two gynecologists, a radiologist and three GPs, as well as X-ray and ultrasound machines. Everything—the clinic building, the medical equipment, the tongue depressors, the television and air-conditioner in the bright waiting room—was paid for by PDVSA. Oil revenues even fed Quintero, who shopped at a subsidized grocery store, part of a chain called Mercal, adjacent to the nucleus. This store sold sugar, rice, milk, cheese and other items for discounts as high as 50 percent; the walls of the Gramoven Mercal were covered in murals that showed a slave breaking his chains. The country had thousands of these stores and, it seemed, a larger number of revolutionary murals.

It was stirring if you did not let your mind linger too long on economics or history.

In the early 1980s, I'd visited Yugoslavia and toured a factory cooperative. The Yugoslav economy revolved around workers' cooperatives, a proud achievement of the country's longtime leader, Josip Broz Tito, who claimed to have found a third way to prosperity that avoided the brutishness of capitalist managers and the dimness of party apparatchiks. At the cooperative I visited, the workers were the owners, or so the pitch went, and all decisions were made democratically by the workers or a council they elected. Shifts, pay and even disciplinary measures were decided by them. Everything was done fairly, and everyone was happy.

It was splendid and unreal, because Yugoslavia's economy was a sort of Ponzi scheme. The cooperatives did not produce goods that people wanted, and behind the scenes, dim-witted apparatchiks were making the big decisions. The country's showcase industrial product, a compact car called the Yugo, was a punch line for late-night comics. The economy held together because Western nations loaned about $20 billion to the Yugoslav government so that it would not fall under the sway of the Soviet Union. The loans were dispersed by Belgrade to cooperatives like the one I visited, and they stayed afloat until the decline of the Soviet Union meant the West no longer needed to subsidize Yugoslavia. The subsequent contraction of the Yugoslav economy helped trigger civil war in the 1990s.

Venezuela's endogenous nuclei, adult-literacy programs and subsidized Mercals were not being kept afloat by loans. They floated on oil. Under Chávez, output was more than 2 million barrels a day, which meant that when prices were $100 a barrel, Venezuela was producing more than $200 million worth of oil every twenty-four hours. Even after deducting the cost of getting the oil out of the ground and shipping it to markets, it was a lot of money for a nation of 26 million souls—not Kuwait levels of drowning-in-oil riches, but higher on a per capita basis than Nigeria or Russia. You didn't need to be a utopian or Marxist to believe it might be possible to reach the goals enunciated by Rafael Ramírez, who served as oil minister and PDVSA president: "To rescue and redistribute petroleum rent to the benefit of the people . . . to transform the terrible imbalances and social inequalities which, paradoxically, are present in one of the countries with the largest oil endowments on the planet." Few governments had made this happen, and by pouring oil money into programs that reminded me of Yugoslavia, I suspected, Venezuela was galloping toward a mirage.

PDVSA's fastest-growing subsidiary was Palmaven, which ran the firm's social programs and was located in an office tower adjacent to the luxury Radisson hotel. It was unusual enough that an oil company had an entire division devoted to good works, but even stranger that the man in charge of the marquee program—the endogenous nuclei—was a navy officer. Captain Rommel Rangel, whose handshake was military strong and whose civilian clothes were pressed to perfection, was not a typical naval *or* corporate man. He was a Chávez supporter who, like the president, had been born into poverty and become disenchanted with neoliberal policies that hollowed out his homeland in the 1980s. Rangel didn't mind that I arrived at his immaculate office on a Friday afternoon, when the city was emptying out for the weekend; he happily talked as it became dark outside, and his fervor was an evangelical's. When I mentioned the oddity of a navy officer running a social program in a petroleum firm, he smiled and responded by turning globalization on its head, Chávez-style. "Economic development is not as important as social change," he said, with the enthusiasm of a man who has just solved a Rubik's Cube. His optimism was admirable. His plans, less so.

Chávez's policies were born of the notion that because neoliberal economics had failed, its opposite would succeed. His embrace of a radical alternative brought to mind Ryszard Kapuściński's description of oil as "the temptation of ease, wealth, strength, fortune, power." Kapuściński meant that oil seduces rulers into believing it is possible to build a new Rome with little difficulty, because money can do anything. The shah of Iran, Kapuściński noted, promised to create a second America in a generation, but succeeded only in paving the way to an oppressive religious regime that, among its many failures, cannot produce enough gasoline for its drivers. Libya's reserves spurred Muammar Qaddafi to a particular brand of change-the-worldism, hinted at by one of his titles, Brotherly Leader and Guide of the Revolution. After taking power in a 1969 coup, Qaddafi aspired to lead the nonaligned world, then scaled back his ambitions to the Arab world, then to Africa only. He planned to develop nuclear weapons and attack American targets. After three decades of failure, and with his country an economic wreck and politically isolated, Qaddafi finally let go of his radical visions.

Chávez's visions of petrograndeur were geographically vast, too. In addition to subsidizing the barrios of Caracas, he was distributing discounted heating oil to poor families in New York, Philadelphia and Boston. Because these donations were intended to embarrass President George W. Bush, who treated Chávez as a grave menace, PDVSA paid for a full-page advertisement in the *New York Times* that boasted, "Venezuela is warming up the holidays in New York." And not just there—Venezuelan oil was distributed at deep discounts throughout Latin America, with a generous portion going to Cuba. Chávez's regime provided financing for a Latin American TV network, Telesur; bought Argentinean bonds when the government in Buenos Aires was reeling; and sent engineers to Bolivia to run gas facilities that were being nationalized. Venezuela's hemispheric outlays were estimated at nearly $9 billion in 2007, which were several times more than the non-military aid doled out by the United States south of its border. Chávez made no secret of his desire to be this century's Bolívar.

The problem was not, as the Bush administration fretted in those days, that Chávez would turn South America into Cuba writ large. That fear overlooked a geopolitical fact, which is that you can rent political friends in this world but you cannot buy them. Chávez's billions in direct aid were no match for the cultural, corporate and political influence Washington retained in the region; American dominion would not be ended with a few years of subsidies from Caracas. The problem was that Venezuela, believing its mirage, could not afford its friends any more the Soviet Union could afford its satellites in Eastern Europe. The Gramoven nucleus was replicated throughout Venezuela, but these cooperatives, created at great cost to PDVSA, provided a fraction of what the country needed in the way of jobs. Caracas was drowning in the usual mix of oil and unemployment. A core feature of the resource curse, as we've seen, is that although the oil industry dominates an economy, it creates few jobs. High-tech refineries can cost billions of dollars to construct, but once they're up and running, perhaps a few hundred workers are needed to monitor them. If you have as much oil per capita as Kuwait, you don't need to worry about real jobs—you can subsidize a life of indolence for everyone in your kingdom. But Venezuela did not have enough oil for that, and the upshot was that its unemployment rate was well into the double digits even during the (relatively) good times. Caracas had a booming business in luxury cars *and* the highest rate of gun violence in the world for cities not at war. The capital's infrastructure, ignored during decades of economic doldrums, continued to be ignored during the boom. A highway to the airport had to be rerouted for months due to a bridge that was in danger of collapsing; what had been an hour-long commute to the airport required three to four hours over a zigzag of back roads.

Chávez was not deterred. He was a true believer in a new economic order that captivated, for a while at least, most Venezuelans. To understand why, I turned on my television.

The day I watched *Aló Presidente*, Chávez was, as usual, a mix of Bill Clinton and Oprah Winfrey. He sat at a desk under a large outdoor tent, dressed in a short-sleeved shirt, talking and joking with an audience of several hundred people who fanned themselves to stay cool in the muggy shade. With a microphone in hand, he walked among the crowd and asked people about their lives, hugging and kissing whoever praised his government, as all did. When he encountered a Cuban doctor—thousands of them provide free medical care in Venezuela in exchange for free oil to Cuba—he waved at the camera and shouted, "Hello, Fidel!"

The show went on for hours, with Chávez extolling his Bolivarian revolution. Bolívar is something of a fetish object for Chávez, who has said he often talks to the great liberator, who has been dead for more than a century. Occasionally, Chávez leaves an empty seat at a table, so that the liberator's spirit has a place to sit. Chávez has changed the country's name to the Bolivarian Republic of Venezuela, and this episode of *Aló Presidente* was broadcast from a PDVSA facility that had been turned into a Bolivarian university, at which admission was open to all who applied. Chávez has also set up Bolivarian circles, local groups with millions of adherents working on behalf of his policies and, crucially, his election campaigns.

"We are not going to rest our bodies or souls until we get rid of the chains around our homeland," he said during the show. "We offer an alternative to those who want a better road, the Bolivarian path. We don't need money from Washington or the IMF. We are not subordinate to their will. We can do it with oil money."

A band played a tune with the refrain "Long live the revolution," and after singing along Chávez embraced the musicians like old friends. His joviality was genuine; Chávez clearly was enjoying himself and drew energy from these people, who were not wealthy and seemed to love their down-to-earth president-host. Chávez eventually returned to his desk, began to sip a fresh cup of espresso and noticed that the time had flown by. "Okay," he said. "It's now three o'clock, which means I've been talking for four hours already. I feel good!"

The program continued for three more hours, during which Chávez warned of the evils of Halloween, took a call from the Venezuelan manager of the Chicago White Sox and announced a raise in doctors' salaries. In certain ways, the show worked. A leader who hopes to fundamentally alter a dysfunctional economy will certainly benefit from personal charm and political popularity; a mild-mannered technocrat would have a hard time imposing the radical changes that would be the preconditions for prosperity in Venezuela or any blighted country that had failed to benefit from its oil.

Yet Chávez's performance had the feel of what Fernando Coronil, a Venezuelan scholar, described as a state limited to "magic performances, not miracles." I understood this more fully when I went to see Chávez in the flesh at Miraflores Palace, his office and residence.

The Miraflores ceremony was part of the great game of our times—the superpower search for steady supplies of energy. China, which didn't import much petroleum until 2000 yet is now the third-largest importer after the United States and Japan, was doing whatever it could to win the friends and resources it needed. In the realm of oil supplies, long-term relationships and contracts are vital. Modest amounts of oil can be bought on the "spot market," which is where countries and companies buy and sell crude for short-term delivery. For example, the cargo of a supertanker can change hands while on the open seas. (In fact, it can change hands several times.) But the amounts of oil that can be purchased in this way are relatively minor. Most of the world's oil is spoken for in long-term contracts that guarantee deliveries from a supplier for several years at least; values are linked to "spot prices," which are market rates that prevail at the time of shipment.

Through its state-owned companies, Beijing hoped to negotiate long-term contracts for Venezuela's crude (as well as Sudan's, Saudi Arabia's and Equatorial Guinea's, among others'). This was a potential threat to America, which was Venezuela's largest oil customer even under Chávez; despite the Bolivarian rhetoric, two-thirds of Venezuela's exports went to America. Altering this balance was a delicate game. Chávez could shout and threaten as much as he wanted—he could even deride America's president as the devil—but actually stopping oil shipments to his large neighbor up north could lead to serious consequences; addicts do not react calmly when separated from their fixes. China knew this and did not want to provoke America, yet everyone understood that some supplies could be acquired without causing World War III. To woo Caracas, China had even agreed to launch a communications satellite on favorable terms. At Miraflores, Chávez was getting ready to break this news to the world.

In a basement conference room the size of a high school theater, the front rows were reserved for Chinese officers and businessmen. On stage, several executives from the China Great Wall Industry Corporation waited for Chávez, who arrived a half hour late, clad in a blue suit, white shirt and red tie. He is built stoutly and has thick facial features that give him the look of a retired yet still-energetic boxer who would be glad to return to the ring. His skin is dark brown, reflecting his mestizo lineage. He fills a room like warm water poured into a cup. Dressed in a suit or uniform, smiling or scowling, he makes an impression.

After the Chinese and Venezuelan anthems were sung, Chávez, standing in front of a portrait of Bolívar, in whose honor the satellite was named, launched into a speech of the sort that was his trademark—presidential streams of consciousness. He congratulated the Chinese for being clever at math and saluted their women for being so beautiful. He thanked the Chinese government for training Venezuelans in satellite technology, saying they were teaching Venezuela "how to fly." As a visual aid, he flapped his arms like wings. He added that the Chinese had learned to fly under "the great Mao Tse-tung," and because Chávez drew inspiration from Mao's one-party, one-truth pedigree, he smiled broadly and exhorted, "Long live the Chinese revolution!"

The Chinese businessmen, as rigorously mercantilist these days as John Rockefeller was in his time, gazed at Chávez. They didn't seem to know whether the desired response was sardonic smiles or clenched fists, but their expressions veered toward the safe harbor of nodding approval. One of them adjusted the volume on his headset—the speech was being translated into Chinese—as Chávez said, "We don't want to earn money out of this. We're not capitalists. This is about the survival of our country and the destruction of capitalism. Capitalists are generating death!"

Yet capitalists were still buying oil from Venezuela, and lots of it. The substance, like water from the glaciers, tends to flow according to a variety of gravitational forces. There is geography (American ports are far closer than China's), technology (American refineries were equipped to process Venezuela's heavier crudes) and, of course, political realities (a cutoff might put Washington into a regime-changing frame of mind). A *presidente* can flap his arms in Caracas and hold his nose at the United Nations and promise to remake his nation, but these are performances. The political or economic miracles that Chávez or any leader in his situation might wish for are, most likely, beyond reach, and have always been so.

A pop quiz:

What is the name of the Venezuelan president who described the backers of globalization at the World Bank as "genocide workers in the pay of economic totalitarianism"?

Which Venezuelan leader, nationalizing the operations of Exxon and other foreign companies, described their corporate activities as "economic oppression"?

Which populist *presidente* poured billions of dollars into social programs, vowing that the wave of oil money washing into the country would be used to create a "Great Venezuela"?

If you answered "Hugo Chávez," you are wrong. The correct answer is Carlos Andrés Pérez, who was president from 1974 to 1979 and dominated Venezuela so thoroughly that he was known by just his initials, CAP. Pérez came to power as Venezuela began gorging on petrodollars in the wake of the 1973 OPEC embargo. For a hallucinatory period, Venezuela had the per capita income of West Germany, the supersonic Concorde flew to Caracas three times a week and in Miami's luxury stores Venezuelans were known as *dame dos*, for "give me two" in Spanish. Pérez was not as virulently anti-American as Chávez but was every bit the populist. He boasted of walking across the entire country during his presidential campaign, visiting every village on foot. Because he assumed the postembargo windfall would be permanent and ever-growing, he authorized billions of dollars in foreign loans to plow ever more money into his Gran Venezuela programs.

This was the economic equivalent of a binge destined to end with the money running out or the bloated corpus of Venezuela being ruined by the windfall. As things turned out, both happened. One of the few people who foresaw this was Juan Pablo Pérez Alfonzo, the former oil minister credited with coming up with the idea of a cartel of producing nations in the 1960s. (Pérez Alfonzo is known as the father of OPEC.) In semiretirement, Pérez Alfonzo told a visiting academic researcher, when Venezuela was afloat on its first oil bonanza, "Don't study OPEC. Study what oil is doing to Venezuela. Ten years from now, twenty years from now, you will see, oil will bring us ruin. . . . It's the devil's excrement."

To understand where Chávez was taking Venezuela, I looked not to the future but to the past. When oil prices collapsed in the 1980s, Venezuela came undone as public debt and national poverty soared. There was no economic safety net, because the influx of oil money had decimated the agricultural and industrial sectors by inflating their costs. They had lost the competitive edge they'd had before the oil boom. As in Saudi Arabia and other oil-exporting countries, more people looked to the government for their sustenance rather than to their own brawn or brains. Yet the government, at times of low oil prices, had little to offer. The country went into a free fall.

Because oil can instigate any absurdity, CAP was brought out of retirement and reelected president in 1988, with a desperate nation hoping he could resummon the prosperity that had existed in his previous reign. Yet he had no more magic tricks or even performances. He bowed to global economic winds and imposed an austerity program that had the short-term effect of making the poor even poorer. Widespread rioting broke out, and as many as three thousand people were killed in the Caracazo, as the 1989 disturbances were called. Among the country's impoverished—and this was now most of the country—the perceived cause of their misery was not oil or debt but the capitalist order. Army officers staged two coups against CAP, and though these uprisings were quashed, the leader of the first one, a charismatic lieutenant colonel, became a national hero for defending the interests of the poor. After more than two years in jail, Lieutenant Colonel Hugo Chávez was set free.

There is a saying that Venezuela does not have good or bad presidents, just presidents who serve at times of high or low oil prices. Chávez, running for president in 1998 as the main political parties all but collapsed from decrepitude, had the great luck of being elected when oil sold for just $12 a barrel. As his presidency got under way, prices began climbing, and six years later, a barrel fetched more than $65, on its way to more than $140. Under his caffeinated direction, Venezuela began a radical spending spree that was similar to the Gran Venezuela effort of a generation earlier, and for economists like José Toro-Hardy, it was akin to watching a car speed toward a wall that it had smacked into not so long ago.

If you don't mind surveillance cameras and ten-foot walls topped with shards of glass, or barking German shepherds and private security guards who glare at all newcomers, the neighborhood of El Country Club is delightful. It is one of the capital's enclaves of wealth and is noted for its namesake, a two-hundred-acre club built in the early

twentieth century for the benefit of American oilmen and their good
friends, the local oligarchy. El Country Club has horse stables, tennis
courts and an eighteen-hole golf course, and is such an untouchable
institution that when the mayor of Caracas proposed confiscating its
land to build low-income housing, Chávez's federal government
advised that this would be an excessive act of Bolivarism.

For the fortunate Venezuelans who reside along the area's spotless
and quiet streets—a universe apart from the chaos in the rest of Caracas—these were wealth-enhancing times. Although Chávez was despised by the upper class, some of whom had already decamped to
luxury condos in Miami, the flood of oil money did not bypass them.
When I visited in 2005, the monthly rent on a three- or four-bedroom
apartment near El Country Club was running at about $7,000. The
providers of luxury goods and services, from late-model BMWs to plastic surgery, were doing a booming business in precincts around the club.

This was where José Toro-Hardy lived, and after I was buzzed
through a locked gate along his street, a maid opened the front door
and walked me through an open-air atrium. The quiet and the green-
ery and the chirping of birds imparted an eco-resort vibe, but Toro-
Hardy, unlike his home, was not at peace. A former director of PDVSA
who was ousted after Chávez came to power, Toro-Hardy had become
a fierce critic of the president's economic policies. His critique rose
above the entitled whinings heard on the shady verandas around El
Country, because Toro-Hardy was the author of scholarly books on oil
economics and considered himself a nationalist whose nation was yet
again being ruined by oil.

His demeanor was nervous, with his eyes darting and his voice
wavering, as if he were a fugitive from the illusion of reality Venezuela
was embracing. His warnings were issued in the manner of a man
imparting what he believed to be a vital yet ignored truth; his pulse and
blood pressure made him a candidate for immediate bed rest.
Venezuela, with the world's highest proportion of beauty queens, was a
nation of obsessives, and Toro-Hardy was obsessed about oil. He led
me into his office and asked me to peruse economic charts as he
searched the Internet for up-to-the-second oil data. One chart told a

notable story. From 1920 to 1980, Venezuela had the strongest growth of any country in South America. For much of that time, oil was not a curse, largely because it was not a dominant force—with oil prices quite low, Venezuela also maintained vigorous farming and industrial sectors. Another chart showed how the petrodollar flood that began in 1973 had upset the country's balance. This is a paradox of windfalls like the one inundating Chávez-era Venezuela; they can distort rather than strengthen national institutions. We've seen this before: as the oil sector grows, farming and manufacturing may contract, unemployment may expand, inflation may rise due to the influx of revenues from oil sales, and the gap between rich and poor may widen. This began in the 1970s, and the latest boom was only papering over the structural problems—it was the euphoria of the bubble.

"This is not something that can be sustained," Toro-Hardy said. "The whole economy depends on government expenditure, but that depends on one factor: oil prices. The laws of economy cannot be violated any more than the laws of gravity. Sooner or later we will have a serious economic crisis."

It was hard to imagine prices returning to the $12-a-barrel lows that had prevailed when Chávez came to power, but even high prices were not a guarantee of high revenues. Like almost every oil exporter, Venezuela struggled to maintain its output. After the mass firing in 2002, PDVSA's output plunged by nearly 50 percent, and not all of the lost ground had been regained. Geology was not helping matters, because Venezuela was running short of the light crude that is easiest to refine. The country has vast deposits of tar sands near the Orinoco River, but converting them into conventional oil is a complex process that involves costly technologies, large volumes of water and natural gas—and it causes severe environmental damage. To get the job done, PDVSA needed the help of foreign firms that were now reluctant to get involved because they had been forced to cede control of oil projects they'd operated before Chávez came to power. Orinoco's heavy oil seemed unlikely to fund the Bolivarian revolution.

Toro-Hardy saw a landscape of problems that centered on dogmatic economic programs implemented not by a government ministry but by an oil company that was having a hard time just pumping oil. He was ready to acknowledge that PDVSA, when Chávez came to power, had needed to be reformed because it had indeed grown aloof from the country. But Chávez's method, firing half the workforce, was akin to destroying a village in order to save it. Extracting oil requires immense amounts of expertise, in the form of engineers who understand the geological profiles of the reservoirs they are drilling into. These experts cannot be replaced like waiters in a restaurant. When large numbers of oil experts left Iran after the 1979 Islamic revolution, production plummeted; Iran's production has inched back, after three decades of war and instability, to just 4 million barrels a day, which is a third less than its prerevolution rate. A similar decline was taking place in Venezuela, where the output was no more than 2.5 million barrels a day when I visited—or a third less than its peak. PDVSA, shorn of its best and brightest, could not do its job as well as it used to, and now it had an additional task to perform.

Chávez did not just order PDVSA to boost its community spending by a few percentage points; he turned the firm into the engine of revolutionary change. PDVSA allotted more to its social projects in 2006—nearly $10 billion—than to its operations ($5.9 billion). In a sense, it became a development agency with oil wells. No other oil company, whether publicly traded or state-owned, spent nearly as much on noncore programs. In Saudi Arabia, Russia and other oil countries, state-owned firms tend to have modest social programs. Their surpluses are transferred to the Treasury and distributed to ministries that chase the holy grail of sustainable development. Usually they fail. You can build colleges, as Saudi Arabia did, but that doesn't mean the degrees will count for much or that jobs will await the graduates.

Chávez was betting, almost literally, that an oil company would succeed where government ministries might not. PDVSA went from one extreme—disassociated from the government it was supposed to serve—to the opposite extreme of taking over the government's duties. I knew that villagers in the Niger Delta would be delighted if Shell or its government-owned partner would provide the education, electricity, medical care and jobs that the negligent and corrupt government did not offer. But it was hard to imagine how oilmen might do better development work than a government's development experts. Oil companies should certainly provide funding and support to official efforts, as well as fight corruption and waste. But replacing a government seemed a doomed concept. As Toro-Hardy said in his exasperated way, "Oil companies should do more, but they should not change their mission. Now, instead of investing in its own projects, PDVSA is investing in housing and social programs. This is very nice, but it's not for an oil company."

In Venezuela, it was as though a well-meaning doctor was using the wrong instruments and wrong procedures to operate on a sick patient. Even during the boom years, signs of failure were ample—price controls on foodstuffs were leading to shortages, and the government was spending so much on subsidies that it was running into deficit problems, which is a striking achievement when large amounts of revenues are being received from oil sales. Chávez's policies, intended to break the resource curse, seemed likely to prolong it. "I am not defending the previous governments," Toro-Hardy said as he walked me out of his private sanctuary. "They did an awful job. But giving away money is not going to solve people's problems. We have a saying here: 'Bread for today and hunger for tomorrow.'"

When oil prices slid back to the double digits, Chávez's popularity began to slide, too. He didn't have as much money to throw at the country's problems. An opposition candidate even won election as mayor of Caracas in 2008. Magic shows can obscure reality but cannot make it disappear.

One day, my journey into the twilight of oil took me to San Gorgonio Pass, in southern California. The mountains around the desert pass rise to more than ten thousand feet, but they are not the most dramatic sight. The pass, one of the breeziest spots in the state, is home to a wind farm that consists of about three thousand turbines spread over five thousand acres. Some turbines are eighty feet high, others rise above three hundred feet, and together they can produce electricity for hundreds of thousands of homes. Set against the blue sky and the brown desert, in rows of rotating white arms that glint in the sun, the turbines have the appearance of futuristic totems waving at us, luring us forward.

The farm is located at the intersection of the I-10 Freeway and Route 62, which leads to a Marine Corps base at Twentynine Palms. This makes San Gorgonio a symbolic as well as a geographic cross-roads. In one direction lies a bastion of American military power that upheld, in the last century, an economic system dependent on fossil fuels. This direction leads, or should lead, to our past. In another direction, the one symbolized by windmills rather than howitzers, lies our future.

Though oil provides fuel for our cars and warmth for our homes, it undermines most countries that possess it and, along with natural gas and coal, poisons the environment. We need to find another way. Because I am hopeful, I have not been speechless when people have asked me, "How do we stop the human, terrestrial and climate damage of fossil fuels?"

I tell friends and strangers about the importance of conservation. I stress the benefits of renewable energy. I note that coal plants are particularly deadly—and that we should build no more of them. Although I haven't raised my own vegetables, I mention the importance of locally grown food and, in the developed world, meals that involve lesser amounts of meat. Of course I emphasize the importance of transparency in oil and gas deals.

This isn't always what my questioners want to hear, though. They want a new answer, something they haven't heard before, a fresh solution to monumental problems that other answers haven't seemed to solve. A new technology, a new . . . something. But the good news, which they haven't understood, is that we already possess most of the answers we need. We have technologies and policies that can, to borrow a phrase from a previous generation, change the world. One of the reasons we face a world melting into violence—and just plain melt-

Wind farm in the San Gorgonio Pass near Palm Springs, California

A worker at the Bibi Heybat oil field in Baku,
Azerbaijan, holds the local currency.

ing—is that for several decades we have refused to act on the answers within reach. Do you remember the solar panels President Jimmy Carter installed on the roof of the White House in 1979? If you don't, that's probably because Carter's call for "the moral equivalent of war" in the realm of energy went nowhere and President Ronald Reagan took down the panels several years later. Yet solar power remains an answer that can help us survive into the twenty-second century.

There are other answers.

Some of this book focused on corruption in countries that have the misfortune to possess large amounts of oil. A remedy is at hand. It's not a complete one, but it could, if enacted in full, return a measure of health to sick economies and polities.

It is known as Publish What You Pay and is being promoted by a nongovernmental organization of the same name. It means what it sounds like. Few companies and governments disclose the contracts they sign or the payments they make and receive. This creates a cloud of confidentiality in which bribes can be paid, sweetheart deals can be struck and billions of dollars can be embezzled. A secret contract is a harbor for crooked executives and politicians.

PWYP would compel companies and governments to publish the financial terms of their contracts. A related though less aggressive campaign that already involves some governments is known as the Extractive Industries Transparency Initiative, and though its goals are similar to PWYP's, it promotes voluntary codes rather than compulsory ones.

A few companies have published some figures, and a handful of nations have publicized a portion of their receipts. But the steps taken so far are extremely small. Today, people in Angola, Equatorial Guinea, Azerbaijan, Turkmenistan, Russia, Iraq, Iran, Saudi Arabia, Sudan and most other energy-rich countries have almost no way of knowing, even though their leaders might have signed on to EITI, how much money is paid to their governments or the terms of payment, and almost no way of confirming that energy revenues go into reliable accounts that are beyond the reach of corrupt leaders. Full transparency would help remedy these problems.

The enforcement of prevailing laws is a remedy, too. There is a piece of philosophy from the frontier days of America: If you are involved in a shootout, you should not have any bullets left at the end. Anti-bribery laws in America and Europe are unevenly enforced. For instance, there has been no sanction against the energy companies that made dubious payments in Equatorial Guinea to President Obiang and his family. Riggs Bank, which helped hide Obiang's money, was forced to pay millions of dollars in fines, although its principal owners, members of the Allbritton family, escaped indictment and were able to sell their bank. Riggs, however, was a niche institution with less clout in Washington than, say, Exxon or Chevron, which have not been indicted or fined for their questionable dealings. There have been some enforcements lately, of course—Halliburton paid substantial fines for its bribery in Nigeria, and Siemens A.G., the German engineering firm, paid fines of $1.6 billion in America and Europe after admitting to a global bribery spree. But much more can be done. Every year U.S. prosecutors issue only a modest number of indictments under the Foreign Corrupt Practices Act. Authorities in Europe and Asia have proven loath to file bribery charges except in the most egregious of cases. The law is a weapon that rarely leaves our holster.

At the risk of sounding even more old-school, I'll mention an additional remedy we need to impose: social values. Even when enforced aggressively, laws alone cannot do everything; they need to be complemented with social pressure that opposes unethical and exploitative profiteering. Ida Tarbell noted this a century ago, in her famous exposé of the extortionary methods used by John D. Rockefeller's Standard Oil Company. Tarbell argued that part of the problem resided in society's ambiguous reaction to Rockefeller's law-shaving, fortune-making success. "There is no cure but in an increasing scorn of unfair play—an increasing sense that a thing won by breaking the rules of the game is not worth the winning," Tarbell wrote. "When the businessman who fights to secure special privileges, to crowd his competitor off the track by other than fair competitive methods, receives the same summary disdainful ostracism by his fellows that the doctor or lawyer who is unprofessional receives . . . we shall have gone a long way toward making commerce a fit pursuit for our young men." Tarbell, who was right one hundred years ago, is right today.

Just as individuals in the developed world need to be partisans of a new ethos, their governments need to encourage good governance in unstable nations that are rich only in resources. Where civic groups and multiparty systems do not exist or are weak, our governments should lobby for them so that grievances can be settled through discussion rather than violence. Democratic governments should not support dictatorships whose oppression foments rebellion. War-crimes trials should be pursued against armies and militias that commit atrocities (as many do in wartime). Development assistance should focus on reducing the reliance on mineral exports like oil.

On paper, these policies have been adopted by democratic and even nondemocratic governments in the developed world, but in practice the situation is quite different. In Azerbaijan, Angola and a barrelful of other dictatorships, support for democratic change has been minimal. On the economic side, development policies pursued by the

World Bank and the IMF have tended to do more harm than good in recent decades, leading to heavy debt loads and industrialization programs that harmed all-important agricultural sectors. We have plenty of answers; resolve is what we lack.

The skeptic will point out that even the best of remedies, administered in the right amounts, cannot turn a curse into a blessing. This is true. There is too much corruption-inducing, economy-deforming, conflict-enhancing, fate-altering value locked up in natural resources like oil. As I've said, only the most stable of democracies, such as Norway, and the tiniest of emirates, such as Abu Dhabi, have avoided the downsides of dependency on a natural resource. They are the outliers. The great middle of resource-rich nations do not have bribery-proof institutions or enough oil to make everyone rich beyond complaint. To exorcise the resource curse, you would almost need to get rid of the resources. I had that wish at times. You cannot navigate the violent creeks of the Niger Delta or visit the contaminated mess of Ecuador's Oriente region without thinking that everything would be better if oil had not been found.

We cannot undo geology, but we can try to make these minerals less valuable over the long term so that afflicted nations might have a chance to reset their priorities. The twilight of oil, after a century in which the resource reshaped the world, will last for years, after all. When we wake up tomorrow, we will still be dependent on petroleum and complicit in the forms of violence—physical, environmental and cultural—that are the consequences of its extraction. The twilight needs to be as short as possible. This is where a fortunate convergence occurs between the answers to global warming, peak oil and the resource curse.

Almost every climate scientist agrees that catastrophic warming will occur if we fail to dramatically reduce our carbon emissions. Changing light bulbs and driving a Prius are just the first, the smallest and the easiest of steps (though only if you can afford a Prius). The good news, according to Princeton University scientist Robert Socolow, is that "humanity already possesses the fundamental scientific, technical and industrial know-how" to solve the climate problem.

Socolow and his colleague Stephen Pacala have introduced the notion of "stabilization wedges"—a set of programs, ranging from conservation efforts to fuel efficiency and renewable energy, that would collectively reduce our emissions to levels that should ward off catastrophic warming. But it won't be easy. New technologies must be proved and scaled up, massive amounts of money must be invested, and lifestyles oriented around cheap fossil fuels must be altered. Socolow and Pacala describe their plan as "a limited set of monumental tasks."

There is a healthy debate over the best solutions, but when scientists shake their heads in grave doubt it's usually not because of problems with wedge 5 or wedge 7. They are most concerned about whether we will do what we can do. The United States is by far the worst offender. It is the world's largest consumer of energy and the second-largest emitter of greenhouse gases (China, due to its reliance on coal, the "dirtiest" of fossil fuels, has nosed ahead in the greenhouse gas category). Even in the Barack Obama era, the United States continues to lag behind Europe in the development of everything from solar panels to carbon trading and mass transit. It's impossible to blame a shortage of ideas. New technologies and Einstein-level genius are not required for new railways and wind farms like the one at San Gorgonio Pass.

My travels in the unhappy precincts of oil intensified my support for the efforts to avert global warming. A world in which the priority is not getting oil but getting *off* oil would be better not only for the atmosphere but also, as I've seen, for the people who live in Nigeria, Iraq, Equatorial Guinea, Russia, Iran and other resource-rich nations. The advent of peak oil is yet another incentive to cut our dependency, because in the years ahead the price will only rise—skyrocket, really—if we fail to arrest our desires for it. If you are concerned about spending too much money on gasoline, just sit back, do nothing and see where those prices are in five or ten years.

If you happen to believe in omens, it's tempting to think that the sky and earth are speaking to us with their shouts of global warming and peak oil, warning us to reduce our consumption of fossil fuels. If we respond, the climate will survive in its present form and oil prices will have a far better chance of staying at reasonable levels. The substance, less lucrative than before, will offer less of a temptation toward bad governance, and it will play a smaller role in our decisions to wage war. After all, why invade country X or support tyrant Y if you can survive without the dark liquid they possess?

These issues were on my mind as I drove into the San Gorgonio Pass and saw its windmills. I would have liked to linger, but another destination called me. I turned onto Route 62 and followed it until I reached Twentynine Palms and its sprawling military facility, which is the base for the marine battalion that toppled the statue of Saddam Hussein. The base, with thousands of marines wandering around in desert uniforms, with rows of beige Humvees and the occasional *whoop-whoop* of helicopters overhead, brought me back to Iraq and everything I had seen there.

I visited Colonel Bryan McCoy, who'd given the order to drop the statue of Saddam Hussein. I noticed a couple of cardboard boxes in his office, half filled; he was packing up before his battalion deployed on a new mission. It was a Friday afternoon, so the battalion's headquarters was emptying out for the weekend. We headed for the officers' club to have a drink with the major who'd driven the first tank into Firdos Square. It was cool and dark inside, where just a few other marines with crew cuts were passing the afternoon in a restaurant that had an amiably dingy feel. We sat near the bar, drinking beer, eating popcorn and bringing ourselves up-to-date on each other's lives.

We talked about Firdos—for these marines it was a small event that, to their surprise, had become a global icon. But mostly we talked about the battles fought on the way to Baghdad. The battalion had lost several of its own in the invasion, including a scout sniper shot dead during an ambush; McCoy was just a few feet away when it happened. The battalion would head back to Iraq in the coming months, because invasion had turned into occupation, and McCoy knew it would lose more troops while shedding the blood of others.

After a few hours it was time to go, so we left the club and blinked into the still-bright sun of the Mojave Desert. After our farewells, I got into my rental car and drove west on Route 62. Soon I was back at San

Gorgonio Pass, looking up at the beckoning windmills. The war in Iraq had not left my mind; nor had the trouble in Nigeria, Ecuador, Russia, Saudi Arabia and other oil-drenched corners of our planet. I realized that in just a few miles I had gone from one vision of the future to another. I knew for sure that the windmills were far more revolutionary than all the toppled statues in the world.

Acknowledgments

I am not alone as a writer. I am surrounded by people who share their time, knowledge and friendship, to the extent that it is misleading for my name to appear alone on the cover. In these final words I wish to thank a cast of generous collaborators.

From the day he plucked my first manuscript from obscurity fourteen years ago, Jonathan Segal has been my editor at Alfred A. Knopf as well as my greatest supporter and friend. Jon not only turns my words into prose, he provides the sort of guidance and stability that are vanishing in the publishing world. Along with Sonny Mehta, editor in chief of Knopf, Jon is a literary treasure. He is truly the best editor and advocate a writer could wish for. Without him, this book would not exist.

For the marathons that are my books I stay in shape by writing articles for the *New York Times Magazine*. Editor in chief Gerald Marzorati has sent me around the globe, allowing me to learn how the world works. Without Paul Tough, who edits my stories at the *Times*, I couldn't have written narratives worthy of publication. I really don't know where I'd be without his help.

It is common for writers to thank their agents, but my esteem for Kathy Robbins exceeds the norm. Counselor, guardian, friend—there is no role or task that Kathy and David Halpern, who works with her, have not fulfilled. It is hard to imagine a writing life without them.

Quite obviously I am able to write because of the special literary biosphere I am fortunate to inhabit. The people I have just mentioned have been constant presences over the years, but for this book I was helped by an additional group of collaborators.

I often worked in foreign countries with interpreters who did much more than translate; they were political and cultural guides who also kept me safe at moments of risk. In Iraq I had the great fortune of teaming up with Thaier Aldaami, who learned English by listening to the songs of Donny and Marie Osmond. I also worked with Salam Pax, a blogger who was reading a Philip K. Dick novel when we met.

In Nigeria I was led by Tony Iyare, who had the ability to not only smooth my passage into the violent Niger Delta but to smile throughout our journey. In Ecuador, where things went slightly more easily, Yury Guerra and Julian Larrea were my linguistic companions. In Venezuela, Leonardo Lameda steered me from Gramoven to Miraflores. In Russia, Anya Masterova was at my side. In Pakistan, Majeed Khan arranged entrée into the halls of power and the corridors of madrassas; his humor was a welcome bonus.

I conducted hundreds of interviews with executives, politicians, dissidents, activists, lawyers, warlords and others. Some of them may not agree with my views, but they were kind enough to talk without preconditions. Those who were generous with their time or facilitated my work include Matthew Simmons, Sadad al-Husseini, John Bennett, Frank Ruddy, Carlos Robles, Gabriel Nguema Lima, Dokubo Asari, King Tom Mercy, Chris Finlayson, Steve Donziger, Donald Moncayo, Marlon Santi, James Giffen, J. Bryan Williams, Colonel Bryan McCoy, Osama Kashmoula, Mohammed Aboush, Dathar Khashab, Captain Tom Hough, Bassim Alim, Ibrahim al-Mugaiteeb, Vagit Alekperov, Bernard Mommer, José Toro-Hardy and Charlie Cooper.

With journalists facing economic challenges, the support of nonprofit and educational institutions is more important than ever. I was fortunate to have a semester-long teaching position at Princeton University's Council of the Humanities, a spell as a Regents' Lecturer at the University of California at Berkeley, and a residency at the Blue Mountain Center in the Adirondacks. All were wonderful respites from the slog of writing as usual.

In much of my travels I had the luck of working with remarkable photographers, including Gilles Peress, Christopher Anderson, Laurent van der Stockt and Thomas Dworzak. This book features pictures by Anderson and Dworzak, as well as Michael Kamber, Ed Kashi, Antonin Kratochvil and Christopher Morris, among others. I have often thought that if my words could be half as powerful as their photos, I would have succeeded at my job.

Friends have pitched in, too. Michael Massing has been a source of enlightenment and advice. Alexander Stille and his son, Sam, have been warm and inspiring hosts. Catherine Talese provided crucial advice on the photos that appear in this book. I received excellent readings on various chapters from Paul Tough, Ken Silverstein, Michael Watts, Owen Matthews, Robert Worth, Michael Massing and Chris Anderson. Chuck Wilson and Dan Kaufman did the hard work of fact-checking. Thanks also to Gary Bass, Anne Kornhauser, Richard Kaye, Ernest Beck, Julie Lasky and Ruti Teitel. And as ever, thanks to my colorful family—mother, siblings, stepmother, nephews, nieces and cousins.

I have saved the best for last. My wife, Alissa Quart, is my love, my friend, my teacher and my partner in mirth. She had the patience to read early drafts of my manuscript and the wisdom to realize when I was heading into a ditch; she shaped this book. The writing process was not easy for me, but I suspect it was even more difficult for her, because an author struggling with a manuscript is an author in pain. I have dedicated this book to her, but more justified is the dedication of my life to her.

Peter Maass
New York City
MAY 2009

APPENDIX A

WORLD CRUDE OIL RESERVES*

SELECTED COUNTRIES

(First figure is barrels of oil in billions; second figure is percent of world total)

Saudi Arabia	264.1	21.0%
Iran	137.6	10.9%
Iraq	115.0	9.1%
Kuwait	101.5	8.1%
Venezuela	99.4	7.9%
United Arab Emirates	97.8	7.8%
Russian Federation	79.0	6.3%
Libya	43.7	3.5%
Kazakhstan	39.8	3.2%
Nigeria	36.2	2.9%
US	30.5	2.4%
Canada**	28.6	2.3%
Qatar	27.3	2.2%
China	15.5	1.2%
Angola	13.5	1.1%
Algeria	12.2	1.0%
Mexico	11.9	0.9%
Norway	7.5	0.6%
Azerbaijan	7.0	0.6%
Sudan	6.7	0.5%
India	5.8	0.5%
Oman	5.6	0.4%
Ecuador	3.8	0.3%
Indonesia	3.7	0.3%
United Kingdom	3.4	0.3%
Gabon	3.2	0.3%
Equatorial Guinea	1.7	0.1%
Brunei	1.1	0.1%
Chad	0.9	0.1%
Total World Reserves:	1,258.0	
OPEC	955.8	76.0%
Former Soviet Union	127.8	10.2%

Source: BP Statistical Review of World Energy 2009

*Proved reserves of oil: Generally taken to be those quantities that geological and engineering indicates with reasonable certainty can be recovered in the future from known reservoirs under existing economic and operating conditions.

**Canadian oil sands: 150.7 billion barrels

APPENDIX B

TOP 15 COUNTRIES

(Figures in barrels per day)

Canada	1,845,000
Mexico	1,092,000
Venezuela	949,000
Saudi Arabia	944,000
Nigeria	860,000
Angola	644,000
Iraq	587,000
Brazil	334,000
Colombia	254,000
Russia	219,000
Algeria	215,000
Ecuador	210,000
Kuwait	181,000
Gabon	108,000
Norway	103,000

Source: Energy Information Administration. Figures for March 2009.

NOTES

Introduction

5 *"nothing to do with oil"*: Quote from Rumsfeld interview with Steve Kroft in November 2002. Available at www.defenselink.mil/transcripts/transcript.aspx?transcriptid=3283.

6 *They are examples:* The resource curse theory gathered force in academic circles after Jeffrey Sachs and Andrew Warner published a National Bureau of Economic Research paper in 1995: "Natural Resource Abundance and Economic Growth." Their paper, which tracked the performance of ninety-five countries between 1970 and 1990, showed a correlation between dependence on natural resources and slow economic growth. "Resource-poor economies often vastly outperform resource-rich economies in economic growth," they concluded. Subsequent research showed variations in the fates of resource-rich nations; the curse was not absolute and was affected by a variety of factors. Some of the best research on the curse has been conducted by Michael Ross of the University of California–Los Angeles, Paul Collier of Oxford University and Terry Lynn Karl of Stanford. Karl's *The Paradox of Plenty: Oil Booms and Petro-States* is a seminal text.

6 *As the graffiti:* Along the Quito–Lago Agrio highway, about forty miles west of Lago Agrio.

6 *Norway was the outlier:* For an incisive look at Norway's unique experience with oil, see Terry Lynn Karl's *The Paradox of Plenty: Oil Booms and Petro-States*, pp. 213–221. "The structures that 'received' Norway's boom could

hardly have been more different from those of the developing countries," Karl wrote. "Oil companies, especially eager to exploit resources outside of OPEC's dominion, did not encounter a poor country, a weak state, undeveloped social forces, or a predatory, authoritarian ruler. . . . Organizing a framework for controlling the oil industry required a high degree of sophistication in planning and administration, which Norway, unlike other oil exporters, possessed in abundance." There is no shortage of reports and articles on the fortunate outcome of the Norwegian experience. Some of the latest include "Thriving Norway Provides an Economics Lesson," by Landon Thomas Jr., *New York Times*, May 14, 2009, and "Frugal Norway Saves for Life After the Boom," by Doug Sanders, *Globe and Mail*, January 31, 2008.

8 *"fill the sight by force"*: Roland Barthes, *Camera Lucida: Reflections on Photography*, p. 91.

1 Scarcity

9 *Simmons is the prosperous founder:* The firm, Simmons & Company International, is based in Houston. Since its 1974 founding, the company says, it has acted as financial adviser on $134 billion in transactions.

11 *Most, if not all:* See *The End of Oil: On the Edge of a Perilous New World*, by Paul Roberts, pp. 48–49.

11 *As a teenager:* Simmons's participation on a high school debate team was chronicled by Mimi Swartz in a *Texas Monthly* article, "The Gospel According to Matthew," published in February 2008.

12 *The geological phenomena: Twilight in the Desert: The Coming Saudi Oil Shock and the World Economy*, by Matthew R. Simmons, pp. 100 and 334.

12 *Daniel Yergin:* See Daniel Yergin's opinion piece "It's Not the End of the Oil Age," *Washington Post*, July 31, 2005.

12 *The price is expected:* For an analysis of the relationship between supply, demand and prices, see the report "Causes and Consequences of the Oil Shock of 2007–2008" by James Hamilton, an economics professor at the University of California, San Diego. Hamilton presented his report at an April 2009 conference sponsored by the Brookings Institution; it has been posted at www.brookings.edu/economics/bpea/~/media/Files/Programs/ES/BPEA/2009_spring_bpea_papers/2009_spring_bpea_hamilton.pdf.

Also, the global recession that began in 2008, though reducing the price of oil in the short term, could make the return of high prices even more dramatic, because some companies have been unable to secure financing for oil-exploration projects. In a story on its Web site, *The Wall Street Journal* noted that "the long-term outlook for oil supply isn't getting any better, which means the prospects for a price spike when demand finally recovers is increasing." See www.blogs.wsj.com/environmentalcapital/2009/04/24/oil-prices-opec-secretary-warns-of-darkening-crude-supply/.

12 *He tells audiences:* See Naimi's speech to an OPEC meeting in Vienna on September 12, 2006. Available at www.saudi-us-relations.org/fact-book/speeches/2006/060912-naimi-vienna.html.

13 *There is an echo of truth:* See the report by the U.S. Senate Permanent Subcommittee on Investigations, "The Role of Market Speculation in Rising Oil and Gas Prices," June 27, 2006.

13 *Simmons reintroduced:* For a lengthier discussion of peak oil, see pages 45–52 in Paul Roberts's *The End of Oil.* For an even lengthier discussion, see Kenneth Deffeyes's *Beyond Oil: The View from Hubbert's Peak.*

14 *Nor can we squeeze:* See "Scraping Bottom," by Robert Kunzig, *National Geographic,* March 2009; "The Costly Compromises of Oil from Sand," *New York Times,* January 7, 2009; "Canada Pays Environmentally for U.S. Oil Thirst," *Washington Post,* May 31, 2006; "An Empire from a Tub of Goo," *Globe and Mail,* January 26, 2008; and Elizabeth Kolbert's "Unconventional Crude," *New Yorker,* November 12, 2007.

15 *Responding to American support:* For an excellent description of the 1973 embargo, see Daniel Yergin's *The Prize: The Epic Quest for Oil, Money, and Power.*

16 *What's known is that:* According to the 2008 BP Statistical Review of World Energy, published in June 2008, the world's proved reserves of oil are 1.2 trillion barrels. Posted at www.bp.com/liveassets/bp_internet/globalbp/globalbp_uk_english/reports_and_publications/statistical_energy_review_2008/STAGING/local_assets/downloads/pdf/statistical_review_of_world_energy_full_review_2008.pdf.

17 *A modern example is Oman:* For more information on Oman's troubles, see "Oman's Oil Yield Long in Decline, Shell Data Show," *New York Times,* April 8, 2004, and "Thirst for Oil Feeds Innovation in Oman," *Washington Post,* August 12, 2008.

17 *Saudi Arabia possesses:* Figures from the BP Statistical Review of World

Energy, 2008. See www.bp.com/productlanding.do?categoryId=6929& contentId=7044622.

17 *Every day, the Saudis provide:* Saudi production varies according to world demand. Aramco claims a production capacity of nearly 12.5 million barrels a day. In recent years, production has ranged between 8 million and 10.5 million barrels a day.

17 *The much-contested reserves:* See the U.S. Geological Service report "Arctic National Wildlife Refuge, 1002 Area, Petroleum Assessment, 1998, Including Economic Analysis." See www.pubs.usgs.gov/fs/fs-0028–01/fs-0028–01.htm

17 *Before visiting the kingdom:* The conference, organized by the Center for Strategic and International Studies, took place on May 17, 2005.

19 *I noted that Royal Dutch/Shell:* See "Can Shell Put Out This Oil Fire?," *BusinessWeek*, May 3, 2004.

20 *This scenario assumes:* The technology is certainly available to significantly increase the use of alternative energy in the medium term. For instance, a key impediment to greater wind power is the absence of transmission lines to distribute wind energy across the country; building a new grid would require large amounts of capital. But combined with rigorous conservation efforts, a shift from fossil fuels is imaginable. The will to do so, as I explain in the conclusion, is another matter.

20 *The end of the suburban: The Long Emergency: Surviving the Converging Catastrophes of the Twenty-First Century* was written by James Howard Kunstler and published by Atlantic Monthly Press in 2005.

20 *"Someday (and perhaps that day will be soon)":* See Simmons, *Twilight in the Desert,* p. 179.

21 *"I can read two hundred papers on neurology":* Quoted in the *Financial Times,* February 27, 2004, in "Saudi Aramco Dismisses Claims over Problems Meeting Rising Global Demands for Oil."

21 *"This is not the first time":* "It's Not the End of the Oil Age," *Washington Post,* July 31, 2005.

2 Plunder

26 *President Teodoro Obiang's:* His full name is Teodoro Obiang Nguema Mbasogo.

26 *Obiang, whose salary:* For salary information, see Ken Silverstein's posting at Harpers.org, www.harpers.org/archive/2006/04/sb-obiang-eg. For details on the $700 million, see the report by the U.S. Senate Permanent Subcommittee on Investigations, "Money Laundering and Foreign Corruption: Enforcement and Effectiveness of the Patriot Act; Case Study Involving Riggs Bank," issued on July 15, 2004. Copy posted at www.hsgac.senate.gov/public/_files/ACF5F8.pdf.

29 *In Ecuador, oil led to:* See "Drilling into Debt," a report pubished in 2005 by Oil Change International. Copy posted at www.priceofoil.org/thepriceofoil/debt-poverty/.

30 *Equatorial Guinea became:* In 1999 the World Bank demanded, in exchange for loans to help build a 665-mile pipeline that would deliver Chad's oil to an export terminal in Cameroon, unprecedented guarantees that Chad's corruption-plagued government would use the revenues for development projects. Almost as soon as the oil began to flow, Chad's government began to flout the agreement, diverting revenues to weapons purchases. In 2008, the World Bank officially withdrew its support for the project, but the withdrawal had no practical effect on the flow of oil and money, as the pipeline was fully operational by then and did not need further support from the bank. See "World Bank Pulls Plug on Chad Oil Pipeline Agreement," by Lesley Wroughton, *Reuters*, September 9, 2008. The bank said in a statement, "Regrettably, it became evident that the arrangements that had underpinned the bank's involvement in the Chad/Cameroon pipeline project were not working."

31 *Equatorial Guinea may be:* My account of the history of Equatorial Guinea is drawn from a number of sources, including Max Liniger-Goumaz's *A l'Aune de la Guinée équatoriale*, Robert Klitgaard's *Tropical Gangsters: One Man's Experience with Development and Decadence in Deepest Africa*, and Adam Roberts's *The Wonga Coup: Guns, Thugs and a Ruthless Determination to Create Mayhem in an Oil-Rich Corner of Africa*.

31 *At Black Beach Prison, Obiang:* See Roberts, *Wonga Coup*, p. 41.

31 *Macias was sentenced to death:* See Roberts' *Wonga Coup*, p. 39.

32 *One of his aides:* See Ken Silverstein, "U.S. Politics in the 'Kuwait of Africa,'" *Nation*, April 4, 2002.

32 *"has just devoured":* See "Where Coup Plots Are Routine, One That Is Not," *New York Times*, March 20, 2004.

33 *At the outset, the American companies:* See IMF report issued in October

1999, "Equatorial Guinea: Recent Economic Developments," p. 18. Copy posted at www.imf.org/external/pubs/ft/scr/1999/cr99113.pdf.

37 *He bought, for $55 million:* See "Teodoro Obiang s'offre un palace volant," *Afrique Centrale*, January 10, 2004, and Roberts, *Wonga Coup*, p. 51.

37 *His indulgences were almost modest:* See "Malibu Bad Neighbor," *L.A. Weekly*, January 18, 2007.

37 *For a weeklong Christmas cruise:* For a dispatch on the cruise, see *New York Daily News*, August 16, 2006.

37 *It should be noted that Teodorin's official salary:* For details on Teodorin Obiang's salary and other financial details, see the written testimony of Arvind Ganesan, of Human Rights Watch, to the U.S. Senate Committee on the Judiciary Subcommittee on Human Rights and the Law, September 24, 2008. Copy posted at www.judiciary.senate.gov/hearings/testimony.cfm?id=3572&wit_id=7452.

38 *More than a decade after:* On malnourishment, see "The Boom That Only Oils the Wheels of Corruption," by Cesar Chalala, *International Herald Tribune*, August 6, 2004.

38 *"The staggering increases on paper":* See Jedrzej George Frynas, "The Oil Boom in Equatorial Guinea," African Affairs, volume 103, number 413, p. 540, published in October 2004.

39 *"We hope this letter finds you well":* Contents of the letter were detailed at the hearing and a copy of the letter was provided to the author. The financial details of Obiang's accounts at Riggs come from the previously cited Senate report, "Money Laundering and Foreign Corruption."

41 *"Sir," Kareri wrote:* A copy of the letter was provided to the author.

43 *The president's playboy son:* See "African Minister Took Cut of Oil Contracts," *Financial Times*, October 25, 2006.

43 *He noted that he was:* According to a copy of the affidavit received by the author. In "African Minister Took Cut of Oil Contracts," *Financial Times* reported on the affidavit.

43 *"by far the most generous":* See previously cited IMF report "Equatorial Guinea: Recent Economic Developments."

43 *And often, in Equatorial Guinea:* See the IMF report "Republic of Equatorial Guinea: 2003 Article IV Consultation—Staff Report," p. 11. Copy posted at www.imf.org/external/pubs/ft/scr/2003/cr03385.pdf.

44 *"a significant earner of income":* See the previously cited Senate Permanent Subcommittee Report on Investigations, p. 50.

44 *When its report was published:* The hearing was held on July 15, 2004.

48 *Those sessions were private:* The meeting took place on April 12, 2006, in Washington, D.C.

48 *A torturer was receiving:* See "Mba's House: Bush Administration Renting Embassy Property from Known Torturer," posted at Harpers.org on October 25, 2006.

3   Rot

55 *Even Senegal:* Haiti and Congo ranked higher than Nigeria in a United Nations survey of human development. As Michael Watts of the University of California at Berkeley noted in "Sweet and Sour," a paper published in 2008 by the Institute of International Studies at UC–Berkeley, "According to former World Bank President Paul Wolfowitz, at least $100 billion of the $600 billion in oil revenues accrued since 1960 have simply 'gone missing.' Nigerian anti-corruption czar Nuhu Ribadu claimed that in 2003 70 percent of the country's oil wealth was stolen or wasted; by 2005 it was 'only' 40 percent. By most conservative estimates, almost $130 billion was lost in capital flight between 1970 and 1996. Over the period 1965–2004, the per capita income fell from $250 to $212 while income distribution deteriorated markedly. Between 1970 and 2000, the number of people subsisting on less than one dollar a day in Nigeria grew from 36 percent to more than 70 percent, from 19 million to a staggering 90 million. Over the last decade GDP per capita and life expectancy have, according to World Bank estimates, both fallen. The Bank put it this way in 2007: 'Per capita GDP in PPP [purchasing power parity] terms fell 40 percent from $1,215 in 1980 to $706 in 2000. Income poverty rose from 28.1 percent to 65.5 percent and other indicators of welfare—notably access to education and health—also declined.' According to the United Nations Development Program, Nigeria ranks in terms of the Human Development Index—a composite measure of life expectancy, income, and educational attainment—number 158, below Haiti and Congo."

55 *The World Bank estimates:* See "Worse Than Iraq?," *Atlantic Monthly*, April 2006.

55 *A few years ago:* See the Human Rights Watch report, "Criminal Politics,"

footnote on p. 49. Copy posted at www.hrw.org/en/reports/2007/
10/08/criminal-politics.

55 *As for the money:* See the International Crisis Group report "Nigeria:
Want in the Midst of Plenty," published July 19, 2006. Posted at www
.crisisgroup.org/home/index.cfm?id=4274.

56 *This is known, in economics:* For more details on the Dutch disease and its
solutions, see Joseph Stiglitz's "We Can Now Cure Dutch Disease," the
Guardian, August 18, 2004, and Christine Ebrahim-Zadeh's "Dutch Dis-
ease: Too Much Wealth Managed Unwisely," *Finance and Development,*
March 2003, as well as Paul Collier's *The Bottom Billion: Why the Poorest
Countries Are Failing and What Can Be Done About It,* pp. 38–40.

56 *"Oil kindles extraordinary emotions and hopes":* Ryszard Kapuściński, *Shah of
Shahs,* p. 35.

57 *His revolt was crushed in twelve:* For a description of Boro's rebellion,
including the use of Shell's boats, see Karl Maier's *This House Has Fallen:
Midnight in Nigeria,* pp. 122–125, and John Ghazvinian's *Untapped: The
Scramble for Africa's Oil,* p. 24.

57 *In 1994, as martial law:* See "Nigeria Crude: A Hanged Man and an Oil-
Fouled Landscape," by Joshua Hammer, *Harper's Magazine,* June 1996.
Hammer notes that Saro-Wiwa's comment was in response to a leaked
memo from the Nigerian military that stated, "Shell operations still
impossible unless ruthless military operations are undertaken for smooth
economic activities to commence . . . Recommendations: Wasting operations during MOSOP [Movement for the Survival of the Ogoni People]
and other gatherings making constant military presence justifiable. Wasting targets cutting across communities and leadership cadres especially
vocal individuals of various groups."

58 *With casually violent ways:* See "Nigeria's Trigger Happy Police," BBC
May 11, 2001. Posted at www.news.bbc.co.uk/2/hi/africa/1322017.stm.

63 *"Dependence on primary commodities":* The report by Collier and Anke
Hoeffler, "Greed and Grievance in Civil War," was published in *Oxford
Economic Papers* 54 (2004), pp. 563–95. Collier's description of the findings is contained in a paper, "Economic Agendas of Civil Wars," which he
presented at a November 30, 2001, meeting in Bonn, Germany, con-
vened by the German Foundation for International Development and
the Federal Ministry for Economic Cooperation and Development.

64 *A dozen people were reported killed:* My description of the fighting in

Tombia comes from a variety of sources, including local residents and the following news stories: "Self-styled Rebel Seeks Independence for Oil-Producing Niger Delta," published on July 16, 2004, by IRIN, the United Nations–affiliated news agency; "Villagers Flee Troops, Militia Fighting Near Nigerian Oil City," by Dulue Mbachu, Associated Press, September 10, 2004; and "Politics of Oil Inflame Age-old Delta Hatreds," by Dudley Althaus, Houston *Chronicle*, December 17, 2004.

68 *According to a joint report:* See "Strategic Gas Plan for Nigeria," a joint report published in February 2004 by the UNDP and World Bank Energy Sector Management Assistance Programme. Posted at www .esmap.org/filez/pubs/58200861713_strategicgasplanfornigeria.pdf.

71 *According to official statistics:* Cited in "Curse of the Black Gold," by Tom O'Neill, *National Geographic*, February 2007.

73 *Throughout the delta:* For a description of bunkering and the amounts of oil involved, see the Human Rights Watch report "The Warri Crisis: Fueling Violence," published December 17, 2003, and posted at www .hrw.org/en/node/12203/section/1. It states, "Theft of crude oil, known as illegal oil bunkering, accounts for perhaps 10 percent of Nigeria's daily production and is a highly organized operation. Governor Ibori has stated that as much as 300,000 bpd [barrels per day] (or 15 percent of production) are lost because of bunkering activities. The major oil companies operating in Nigeria have stated that this is likely an overestimate; for the whole Niger Delta, illegal oil bunkering probably reaches a maximum 150,000 or 200,000 bpd. But these figures also fluctuate significantly, responding to periodic efforts to police the riverine areas more effectively. There are other claims that the theft of oil is greatly underreported, reaching more than 250 million barrels for the year 2002 (that is, more than 650,000 bpd). The illegally bunkered oil is sold to refineries in Nigeria, in nearby West African states (including Côte d'Ivoire and Cameroon), or further afield."

73 *Two navy admirals:* See "Nigerian Admirals Pay the Price for Stealing Captured Oil Tanker," London *Times*, January 8, 2005.

75 *Shell faced a public relations disaster:* In 2009, Shell agreed to pay $15.5 million to settle a lawsuit filed by Saro-Wiwa's son and nine other plaintiffs who accused the company of complicity in a variety of human-rights abuses, including the arrest and execution of Saro-Wiwa. In the settlement, which was reached just before the trial was to begin in a New York

court, Shell admitted no wrongdoing and called the payment "a humanitarian gesture." The plaintiffs said in a statement, "We are gratified that Shell has agreed to atone for its actions." Their statement is posted at www.wiwavshell.org/documents/Wiwa_v_Shell_Statement_of_Plaintiffs .pdf. See also "Shell Pays Out $15.5m Over Saro-Wiwa Killing," *The Guardian*, June 9, 2009.

78 *The report was remarkable:* Shell has not published an official version of the report, but the leaked draft is widely available on the Internet. National Public Radio has posted a copy at www.npr.org/documents/ 2005/aug/shell_wac_report.pdf.

80 *In MEND's first months:* See "Nigerian Militants Free Italian, Hold Three Other Foreigners," Agence France Presse, January 18, 2007; "In Nigeria's Violent Delta, Hostage Negotiators Thrive," by Chip Cummins, *The Wall Street Journal*, June 7, 2007; "The Risk Premium," by Mimi Swartz, *Texas Monthly*, June 2008; and "Blood Oil," by Sebastian Junger, *Vanity Fair*, February 2007.

4 Contamination

83 *It is an irony:* For an excellent review of the link between oil roads and deforestation, see "Causes and Consequences of Deforestation in Ecuador," published in 2001 by the Center for the Investigation of Tropical Rainforests. The study notes, "Since the early 1970s about 30 percent of the Ecuadorian Amazon has been deforested and/or polluted and entire indigenous cultures, such as the Cofan and Huaorani, have been placed in danger of extinction as a result of the oil industry and accelerated colonization facilitated by the oil roads." Copy posted at www.rainforestinfo .org.au/projects/jefferson.htm.

84 *More than 18 billion gallons:* Texaco, which is now owned by Chevron, has an extensive Web site that offers the company's account of what happened; it's located at www.chevron.com/ecuador/. The company's principal opponents have their own Web site, located at www.chevrontoxico .com.

87 *The first offering from Texaco:* See "Tribe Members Didn't Resist Gifts of Food, Fuel," in *Newsday*, May 22, 2005.

87 *"complete autonomy":* A copy of the affidavit was provided to the author.

See also "Ecuadoreans Want Texaco to Clear Toxic Residue," *New York Times*, February 1, 1998.

88 *Thanks to oil:* See "Drilling into Debt," published in 2005 by Oil Change International.

88 *Instead of investing:* The default in 1999 was for reasons of poverty—Ecuador's strapped government did not have the funds to meet its debt obligations. In 2008, the problem wasn't money as much as law and politics—President Rafael Correa described the debts as "immoral and illegitimate" because, his administration said, they were negotiated on unfair terms and without proper authorization. See "Correa Defaults on Ecuador Bonds, Seeks Restructuring," by Stephan Kueffner, *Bloomberg*, December 12, 2008, and "Ecuador's Debt Default: Exposing a Gap in the Global Financial Architecture," by Neil Watkins and Sarah Anderson, *Foreign Policy in Focus*, December 15, 2008.

90 *Rockefeller then returned:* See Yergin, *The Prize*, p. 109.

90 *Its appeals delayed payment:* The *New York Times* reported on the ruling in a June 26, 2008, story, "Damages Cut Against Exxon in Valdez Case." For the number of plaintiffs who have died since the spills, see "Exxon Valdez Decision Expected in the Next Four Weeks," *Alaska Daily News*, June 1, 2008.

92 *In 2008, two Chevron lawyers:* See "Chevron Lawyers Indicted in Connection with Ecuador Case," by Dan Slater, *Wall Street Journal* online, September 15, 2008. Posted at www.blogs.wsj.com/law/2008/09/15/chevron-lawyers-indicted-in-connection-with-ecuador-case/.

96 *The accusation made national headlines:* See "Woes Mount for Oil Firms in Ecuador," by Kelly Hearn, *Christian Science Monitor*, February 9, 2006.

98 *And the people of Sarayaku were ready:* For a useful overview of the Sarayaku standoff, see "The New Amazon," by Marisa Handler, *Orion*, January 2005.

5   Fear

101 *"Ordinary people":* Stanley Milgram, *Obedience to Authority*, p. 6.

101 *"He steals money from California":* See "Word for Word/Energy Hogs," *New York Times*, June 13, 2004.

106 *James Giffen was the son:* An in-depth description of Giffen's career and

deal making is contained in Steve LeVine's excellent book *The Oil and the Glory*. Portions of my account are drawn from LeVine's work.

107 *According to the Justice Department's indictment:* The U.S. Southern District Prosecutor posted an announcement of the indictment of Giffen and J. Bryan Williams at www.usdoj.gov/usao/nys/pressreleases/April03/giffenwilliams.pdf.

111 *As a convicted felon:* Seymour Hersh wrote a lengthy story, "The Price of Oil" (*New Yorker*, July 9, 2001), that delved into the activities of Williams and Giffen, who each told me they disagreed with the article's portrayal.

111 *"some of the payments":* See "Manhattan Judge Rules on Pre-trial Motions in 'Kazakhgate' Case," by Marlena Telvick, July 9, 2004, published by International Freedom Network. Copy posted at www.ifn.org.uk/article.php?sid=6.

112 *"Bullshit":* See Anne Applebaum, "Fond Memories of Stalin," *Slate*, June 5, 2000.

114 *Tantalizingly, hotel residents could see:* My description of the Intourist draws on interviews with oil executives who were there in the 1990s. I have also drawn from the work of journalists Thomas Goltz and Steve LeVine, who reported from Baku in the 1990s, as well as "Azerbaijan's Riches Alter the Chessboard," by Dan Morgan and David B. Ottoway, *Washington Post*, October 4, 1998.

117 *This was the gala evening:* The conference was organized by Cambridge Energy Research Associates, a consulting firm headed by Daniel Yergin.

117 *He was rewarded:* See "For Leading Exxon to Its Riches, $144,573 a Day," *New York Times*, April 15, 2006.

6   Greed

123 *American oil firms and executives:* After the start of World War II, Texaco secretly shipped oil to Germany via Colombia. The firm's president, Torkild Rieber, even traveled to Berlin to meet Hermann Goering and agreed to convey a proposal for the surrender of Britain. Although President Franklin D. Roosevelt told Rieber to cease all contact with the Germans, Texaco secretly paid the salary and expenses of a German agent

who came to New York to persuade American businessmen not to supply Britain with weapons and other materiel. The German was even given an office in Texaco's headquarters in the Chrysler building. Once Texaco's ties to Hitler's regime were disclosed, the firm's share price tumbled and Rieber was forced to resign. See Anthony Sampson's *The Seven Sisters: The Great Oil Companies and the World They Shaped*, pp. 78–83. As Sampson wrote of the oil scandals in the wartime era, "It is not necessary to see these three scandals as evidence of any special moral turpitude on the part of the oil leaders: they were brigands of their time, trying to extend a greedy international industry across the barriers of war. They were men who did not know when to stop, and there was very little to stop them. But their ruthlessness and autocracy did reveal very sharply the basic uncontrollability of oil, and the ability of the industry to defy national governments."

123 *As Interior Secretary Harold Ickes:* He wrote the remark in one of his diaries. See Sampson, *The Seven Sisters*, pp. 94–95.

123 *Cheating continues in America:* For example, in 2008 the Interior Department's inspector general issued reports that accused more than a dozen current and former officials in the department's royalty-collecting service of engaging in drug use and illicit sex with employees of energy firms, as well as accepting meals, ski trips, sports tickets and golf outings from them. The report said the acceptance of banned gratuities occurred "with prodigious frequency." See *New York Times*, "Sex, Drug Use and Graft Cited in Interior Department," by Charlie Savage, September 10, 2008, and *Washington Post*, "Report Says Oil Agency Ran Amok," by Derek Kravitz and Mary Pat Flaherty, September 11, 2008.

123 *On occasion these missing links:* Details of the scheme are drawn from a number of published sources, including "U.S. Targets Overseas Bribery," by ProPublica and PBS's Frontline, September 9, 2008, and "Out of Africa: In Halliburton Nigeria Probe, a Search for Bribes to a Dictator," by Russell Gold and Charles Fleming, *Wall Street Journal*, September 29, 2004.

124 *In 2009, Halliburton admitted:* See "Halliburton, KBR Settle Bribery Allegations," by Zachary A. Goldfarb, *Washington Post*, February 12, 2009.

124 *Tesler has been indicted:* For Tesler's indictment, see the Department of Jus-

tice announcement posted at www.usdoj.gov/opa/pr/2009/March/09-crm -192.html.

127 *The Dodges filed a lawsuit:* Michael Skapinker wrote about *Dodge v. Ford Motor Co.* in "Fair Shares?," *Financial Times*, June 11, 2005. In an informative exchange, several distinguished law professors had an online debate about the meaning of *Dodge v. Ford;* see www.businessassociationsblog .com/lawandbusiness/comments/does_dodge_v_ford_motor_co_remain _canon/.

127 *Milton Friedman championed:* The article was published in the *New York Times Magazine* on September 13, 1970.

129 *In 1991 Condoleezza Rice joined the Chevron board:* See *Condoleezza Rice: An American Life*, by Elisabeth Bumiller, pp. 109–10.

130 *"Today's energy industry earnings":* *New York Times*, January 19, 2006.

130 *The same year, without any mention:* See Justin Fox, "No More Gushers for ExxonMobil," *Time*, May 31, 2007.

131 *"Would ExxonMobil be willing":* *Today*, May 3, 2006.

132 *"left the church":* Darcy Frey, "How Green Is BP?," *New York Times Magazine*, December 8, 2002.

132 *In 2005, a BP refinery in Texas:* See "Faults at BP Led to One of Worst US Industrial Disasters," *Financial Times*, December 18, 2006, and "BP Paints Grim Picture of Texas Refinery Before Blast," *Financial Times*, March 19, 2007.

133 *"For a company that claims":* See "Behind the Spin, the Oil Giants Are More Dangerous Than Ever," *Guardian*, June 13, 2006.

133 *"What we stand to gain":* See "U.S. Accuses BP of Manipulating Price of Propane," *Wall Street Journal*, June 29, 2006.

133 *As the presses rolled:* See "The Tragic Departure of a Gay CEO," *Newsweek*, May 3, 2007.

133 *"Corporations have to be responsive":* See "Five Who Laid the Groundwork for Historic Spike in Oil Market," *Wall Street Journal*, December 20, 2005.

134 *Until the 1970s:* See "As Oil Giants Lose Influence, Supply Drops," by Jad Mouawad, *New York Times*, August 19, 2008.

7 Desire

139 *In a famous meeting:* The deadly purge has been widely reported. See Neil MacFarquhar, "Saddam Hussein Had Oppressed Iraq for More Than 30 Years," *New York Times,* December 29, 2006, and Bay Fang, "When Saddam Ruled the Day," *U.S. News & World Report,* July 11, 2004.

139 *The United States even supplied Iraq:* Reported by Michael Dobbs, "U.S. Had Key Role in Iraq Buildup," *Washington Post,* December 30, 2002.

139 *Donald Rumsfeld, a pharmaceutical:* See Christopher Marquis, "Rumsfeld Made Iraq Overture in '84 Despite Chemical Raids," *New York Times,* December 23, 2003.

139 *"There were indications":* See Joost R. Hiltermann, *A Poisonous Affair: America, Iraq and the Gassing of Halabja,* p. 7.

140 *"our vulnerable friend Saudi Arabia":* George H. W. Bush and Brent Scowcroft, *A World Transformed,* p. 303.

140 *"They won't stop here":* Ibid., p. 319.

140 *"[Saddam] has clearly done":* Ibid., p. 323.

142 *As the writer:* Stephen Kinzer, *All the Shah's Men: An American Coup and the Roots of Middle East Terror,* p. 2. Much of my description of Mossadegh and the coup comes from Kinzer's invaluable book.

142 *Even British foreign secretary:* Kinzer, *All the Shah's Men,* p. 68

142 *"Ever since Churchill":* Anthony Sampson, *The Seven Sisters: The Great Oil Companies and the World They Shaped,* p. 137.

144 *As Roosevelt later wrote:* Kinzer, *All the Shah's Men,* p. 173.

144 *Roosevelt had shown:* Ibid., pp. 179–80.

144 *The shah, returning home:* Ibid., p. 191.

145 *A month after Iraq's invasion:* The speech was made on September 11, 1990.

146 *Americans sensed this:* See Christopher Layne and Ted Galen Carpenter, "Time for Congress to Vote on the Issue of War in the Gulf," Cato Institute Foreign Policy Briefing No. 5, December 14, 1990. Layne and Carpenter cite several opinion polls, including one published by the *Los Angeles Times,* in which 53 percent of the respondents opposed going to war against Iraq. Posted at www.cato.org/pubs/fpbriefs/fpb-005.html.

146 *Nayirah, a teenage Kuwaiti girl:* The hearing, held on October 10, 1990, was organized by the Congressional Human Rights Caucus, chaired by Democrat Tom Lantos and Republican John Porter.

146 *She tearfully recounted:* My account of the Nayirah saga is taken from a number of sources, including Kathleen Hall Jamieson and Paul Waldman, *The Press Effect*, pp. 16–20; John R. MacArthur, "Remember Nayirah, Witness for Kuwait?," *New York Times*, January 6, 1992; Arthur E. Rowse, "How to Build Support for War," *Columbia Journalism Review*, September/October 1992; and "When Contemplating War, Beware of Babies in Incubators," *Christian Science Monitor*, September 6, 2002.

147 *Bush reinforced the theme:* The speech, on October 28, 1990, was made at Hickam Air Force Base, adjacent to Pearl Harbor.

147 *In the debate:* Micah L. Sifry and Christopher Cerf, eds., *The Iraq War Reader* (New York: Simon & Schuster, 2003), p. 135.

158 *Chevron was one of the buyers:* "Chevron to Pay $30 Million to Settle Kickback Charges," *New York Times*, November 15, 2007.

158 *"It is tempting":* "The War in Iraq Is Distracting the West from the Looming Crisis in Saudi Arabia," Anthony Sampson, *Independent*, May 22, 2004.

159 *"This is the guy that tried to kill my dad":* Bush made the remark at a fundraiser in Texas on September 26, 2002.

163 *As Saddam's regime fell apart:* There are a number of accounts of the killing that day; the details remain murky. I have drawn on many articles, including several stories written by reporters for Knight-Ridder who were in Najaf at the time, as well as an account from *Newsweek* ("Murder at the Mosque," by Joshua Hammer, May 19, 2003) and *The New Yorker* ("The Uprising," by Jon Lee Anderson, May 3, 2004).

160 *"Should all his ambitions":* A transcript of Cheney's speech is posted at www.georgewbush-whitehouse.archives.gov/news/releases/2002/08/20020826.html

163 *"anesthetizes thought, blurs vision, corrupts":* Ryszard Kapuściński, *Shah of Shahs*, p. 35.

163 *One version of events:* Sadr's involvement was mentioned in a report by an Iraqi judge who investigated the killing. "Take him away and kill him in your own special way," Sadr said, according to the judge's account. See "Sword of the Shia," by Jeffrey Bartholet, *Newsweek*, December 4, 2006.

164 *"Oh, occupier":* The slogan was mentioned in Anthony Shadid's August 30, 2005, story in the *Washington Post*, "Sadr's Disciples Rise Again to Play Pivotal Role in Iraq."

164 *About $200 million:* The $200 million figure comes from the *New York Times* story "Iraq Insurgency Runs on Stolen Oil Profits," March 16,

2008, and "Billions in Oil Missing in Iraq, U.S. Study Says," *New York Times*, May 12, 2007.

164 *Until the Iraqi army:* "Oil, Politics and Bloodshed Corrupt an Iraqi City," *New York Times*, June 13, 2006.

8 Alienation

165 *On a warm spring evening:* The reception, held on May 16, 2005, was sponsored by the U.S.–Saudi Arabian Business Council.

166 *Naimi was born in 1935:* Details of Naimi's early life come from John Lawton's "Naimi: 'I Hope to Tell Him "Objective Accomplished,"'" published in *Saudi Aramco World Magazine*, May/June 1984, and "The Arabs," by David Lamb (New York: Random House, 1987), pp. 276–78.

167 *The country's founder:* Dilip Hiro, *The Essential Middle East: A Comprehensive Guide* (New York: Basic Books, 2003), p. 2. There are many estimates on the number of wives; Hiro's number is seventeen, but there appears to be no consensus.

167 *He finally arranged:* For a colorful account of the negotiations, see Daniel Yergin's *The Prize: The Epic Quest for Oil, Money and Power*, pp. 289–92.

167 *"The oil concession":* Madawi al-Rasheed, *A History of Saudi Arabia*, p. 93.

167 *After news of the discovery:* See Wallace Stegner's *Discovery!: The Search for Arabian Oil*, an excerpt of which, detailing the moment of oil being found, was published in *Saudi Aramco World* magazine in January/February 1969; and "Well Done, Well Seven," by Mary Norton, in *Saudi Aramco World* magazine, May/June 1988.

167 *The only Americans:* Anthony Sampson, *The Seven Sisters: The Great Oil Companies and the World They Shaped*, p. 91.

168 *"What does it concern them":* Abdelrahman Munif, *Cities of Salt*, p. 29.

169 *"The moment has come":* Yergin, *The Prize*, p. 606.

169 *"petrodollars actually sever the very link":* Terry Lynn Karl and Ian Gary, "The Global Record," *Foreign Policy in Focus*, PetroPolitics Special Report, January 2004.

170 *He died in 1953:* See Rachel Bronson, *Thicker Than Oil: America's Uneasy Partnership with Saudi Arabia*, p. 75.

170 *Long before Hollywood A-listers:* See "Abramovich Spending Goes Sky-High on Flying Palace," London *Times*, August 22, 2004, and Joe Havely, "Air Force One: The Flying White House," CNN, February 15, 2002.

170 *"a jungle inhabited by beasts of prey"*: The cables have been posted on the Web site of the Campaign Against Arms Trade, at www.caat.org.uk/issues/saudi-bribery.php. The Guardian newspaper has also published them at www.guardian.co.uk/world/2007/jun/07/bae18.

171 *"misused or got corrupted with $50 billion"*: The interview was broadcast in a 2001 PBS documentary, *Looking for Answers*. A transcript of the interview was posted at www.pbs.org/wgbh/pages/frontline/shows/terrorism/interviews/bandar.html.

171 *The prince, whose vacation compound*: Kirk Johnson, "A $135 Million Home, but If You Have to Ask . . . ," *New York Times*, July 2, 2007.

171 *A former fighter pilot*: See David Leigh and Rob Evans, "The Bandar Cover-Up: Who Knew What, and When?," *Guardian*, June 9, 2007, and Nelson D. Schwartz and Lowell Bergman, "Payload: Taking Aim at Corporate Bribery," *New York Times*, November 25, 2007.

171 *Even the advent*: The television protest, as well as the introduction of women's education, is described by Peter W. Wilson and Douglas F. Graham in *Saudi Arabia: The Coming Storm*. London: M.E. Sharpe, 1994, p. 55.

171 *The revolt of the alienated*: My account of the siege is drawn from a number of sources, foremost among them Lawrence Wright, *The Looming Tower: Al-Qaeda and the Road to 9/11*, pp. 88–94.

172 *In the final battle*: For a description of the use of CB gas, as well as the executions, see Yaroslav Trofimov, *The Siege of Mecca: The 1979 Uprising at Islam's Holiest Shrine*, pp. 191–92 and 238–40.

172 *An American think tank*: See "The Saudi Connection: How Billions in Oil Money Spawned a Global Terror Network," *U.S. News & World Report*, December 9, 2003. The $70 billion estimate came from Alex Alexiev of the conservative organization Center for Security Policy. On June 26, 2003, Alexiev cited the estimate in testimony before the Senate Subcommittee on Terrorism, Technology and Homeland Security; a transcript is posted at www.kyl.senate.gov/legis_center/subdocs/sc062603_alexiev.pdf.

172 *With a population*: For Saudi funding as the proportion of the faith's costs, see Lawrence Wright, *The Looming Tower*, p. 149.

173 *The oil boom enriched*: The biographies of Mohammed bin Laden and his seventeenth son are well known by now. Some of the latest sources I drew on include Wright's *Looming Tower* and Steve Coll's *The Bin Ladens: An Arabian Family in the American Century*, as well as Coll's earlier *Ghost*

Wars: The Secret History of the CIA, Afghanistan, and bin Laden, from the Soviet Invasion to September 10, 2001.

173 *Although his wealth:* See Coll, *The Bin Ladens,* pp. 351–52.

174 *The Sudanese all but fleeced him:* See Wright, *Looming Tower,* pp. 196–97 and 222–23.

174 *At times, bin Laden and his followers:* Ibid., p. 248.

174 *"The Saudi Arabian government spares no effort":* Amnesty International, "Saudi Arabia: A Secret State of Suffering," March 2000. Posted at www.amnesty.org/en/library/info/MDE23/001/2000.

177 *More than 30 percent:* Employment statistics issued by the Saudi government are regarded as unreliable, downplaying the actual numbers of the jobless. The estimates I use come from an article by Eric Rouleau in the July/August 2002 issue of *Foreign Affairs* ("Trouble in the Kingdom"). Later estimates have not varied significantly from the figures used by Rouleau.

179 *The crucial thing:* See "Petroleum, Poverty and Security," an excellent report issued by Chatham House in June 2005. The report notes, "Petroleum output per head is a primary measure of petroleum resource flow, in the same way that GDP per capita is a measure of economic wealth." Copy posted at www.chathamhouse.org.uk/files/3254_bppetroleum.pdf.

180 *In the early 1980s:* See "Leisure Class to Working Class in Saudi Arabia," by Neil MacFarquhar, *New York Times,* August 26, 2001.

181 *The rhythm of life:* For population figures, see "Young and Restless," by Afshin Molavi, *Smithsonian,* April 2006. The story notes that not only are 75 percent of the population under thirty years of age, 60 percent are under twenty-one and more than one in three is under fourteen.

182 *Addled on illicit drugs:* See Josh Martin, "Arab Traffic Jam," *The Middle East,* March 1, 2005.

182 *I was not surprised:* The URL is www.youtube.com/watch?v=PJup NDIKkEk.

9 Empire

187 *By 2004, Moscow counted more billionaires:* "Moscow Overtakes New York as the Billionaires' Capital of the World," *Independent,* May 14, 2004.

187 *The local media:* "It Isn't Magic: Putin Opponents Vanish from TV," *New York Times,* June 3, 2008.

187 *As democracy shrank:* On life expectancy, see Nicholas Eberstadt, "Rising Ambitions, Sinking Population," *New York Times,* October 25, 2008.

188 *In the late 1950s:* See Yegor Gaidar, "The Soviet Collapse," published by the American Enterprise Institute, April 19, 2007, as well as Gaidar's *Collapse of an Empire,* chapter 3.

188 *"Without the discovery":* Stephen Kotkin, *Armageddon Averted: The Soviet Collapse, 1970–2000,* p. 15.

188 *But the impact:* See Peter Schweizer, *Victory: The Reagan Administration's Secret Strategy That Hastened the Collapse of the Soviet Union,* pp. 217–20. Also see Marshall I. Goldman, *Petrostate: Putin, Power, and the New Russia,* pp. 49–54.

189 *At the end of 1985:* For a description of the Saudi production increase, see Yergin's *The Prize,* pp. 748–751.

189 *"Gorbachev's incipient perestroika":* Stephen Kotkin, "What Is to Be Done?" *Financial Times,* March 5, 2004.

189 *According to Yegor Gaidar:* Quoted from Gaidar's "The Soviet Collapse." Gaidar has noted that oil revenues propped up the Soviet economy for decades. In *Collapse of an Empire,* he wrote, "The hard currency from oil exports stopped the growing food supply crisis, increased the import of equipment and consumer goods, ensured a financial base for the arms race and the achievement of nuclear parity with the United States, and permitted the realization of such risky foreign policy actions as the war in Afghanistan"; see p. 102.

190 *"Nothing is free":* Paul Klebnikov, *Godfather of the Kremlin: The Decline of Russia in the Age of Gangster Capitalism,* p. 194.

192 *Roman Abramovich, an oil multibillionaire:* Details on Abramovich's acquisitions are from "Abramovich Purchases Equal of Air Force One," *St. Petersburg Times,* May 25, 2004; "A Roman Retreat," *Time,* November 24, 2003; and "I'm No Napoleon, Says Abramovich," *Daily Telegraph,* July 6, 2003.

193 *A few days earlier:* See "Advertisers Doubt NFQ Rival Killed Goldman," *Moscow Times,* April 14, 2004.

193 *Putin reportedly amassed:* See Luke Harding, "Putin, the Kremlin Power Struggle and the $40bn fortune," *Guardian,* December 21, 2007.

194 *Since Putin had come to office:* See Michael A. McFaul and Kathryn Stoner-Weiss, "The Myth of the Authoritarian Model: How Putin's Crackdown Holds Russia Back," *Foreign Affairs,* January/February 2008.

195 *Khodorkovsky was also challenging Putin:* See Marshall I. Goldman, *Petrostate: Putin, Power and the New Russia*, pp. 111–13.

195 *When a Chinese firm tried:* See "Chinese Company Drops Bid to Buy U.S. Oil Concern," *New York Times*, August 3, 2005.

196 *Under the post-Communist rule:* There are a number of excellent books about this period in Russian economic history. One of the best is *Sale of the Century: Russia's Wild Ride from Communism to Capitalism*, by Chrystia Freeland. See also *The Oligarchs: Wealth and Power in the New Russia*, by David E. Hoffman, and Paul Klebnikov's *Godfather of the Kremlin*.

197 *"Putin arrived on the scene":* See Michael McFaul and Kathryn Stoner-Weiss, "The Myth of Putin's Success," *International Herald Tribune*, December 13, 2007.

198 *Not long afterward:* See "Russia Plans to Sell Bonds in 2010, Seeks Loans from World Bank," by Paul Abelsky, *Bloomberg*, April 27, 2009. The story, attributing its information to Finance Minister Alexei Kudrin, stated that "the country's Reserve Fund, one of its two sovereign wealth funds, may be exhausted by the end of this year."

198 *Russia's megarich:* See "Up in Smoke," *Forbes*, March 30, 2009. The magazine reported that in the wake of the global recession, Moscow, which in 2008 had seventy-four billionaires to New York's seventy-one, now had only twenty-seven to New York's fifty-five.

198 *A rash of street protests:* See "Thousands Protest Across Russia," the BBC, January 31, 2009, posted at www.news.bbc.co.uk/2/hi/europe/7862370.stm. The story noted, "Protests on such a large scale were unthinkable just a few months ago as the economy boomed with record high oil prices and as the Kremlin tightened its grip over almost all aspects of society, the BBC's Richard Galpin in Moscow says. But now with the economy in deep trouble, there is real fear amongst ordinary people about what the future will hold."

198 *"The state has become":* See Andrei Illarionov, "Russia Inc.," *New York Times*, February 4, 2006.

10   Mirage

199 *In the halls of American power:* Chávez made the speech on September 20, 2006.

199 *"rich countries with poor people":* See Joseph Stiglitz, "We Can Now Cure Dutch Disease," *Guardian*, August 18, 2004.

202 *But like the foreign companies:* See David Luhnow and Peter Millard, "As Global Demand Tightens, A Big Producer Has Own Agenda," *Wall Street Journal*, August 1, 2006, and Natalie Obiko Pearson, "Chávez's Largesse Puts Strain on Venezuela's State Oil Company as Exports to U.S. Decline," Associated Press, March 27, 2007.

202 *Chávez proceeded to turn:* The statistics and rationale for Chávez's move are explained by one of his top oil advisers, Bernard Mommer, in "Subversive Oil," a chapter Mommer wrote for the book *Venezuelan Politics in the Chávez Era*, Steve Ellner and Daniel Hellinger, eds. (Lynne Rienner, 2004). A useful overview of Chávez's place in Venezuela's petrohistory is contained in Michael Shifter's "In Search of Hugo Chávez," *Foreign Affairs*, May/June 2006.

204 *The subsequent contraction:* See my book *Love Thy Neighbor: A Story of War* (Knopf, 1996).

204 *Under Chávez, output was more than:* There are contradictory statistics on Venezuela's oil production. The government claims a daily production of 3.3 million barrels, but figures from OPEC and the International Energy Agency show output close to 2.4 million barrels a day. See "Venezuela's Oil-Based Economy," Council on Foreign Relations, June 27, 2008, copy posted at http://www.cfr.org/publication/12089/, and Rachel Jones, "Venezuela Seeks Investment from Big Oil," Associated Press, January 15, 2009.

204 *"To rescue and redistribute petroleum rent":* "A National, Popular and Revolutionary Oil Policy for Venezuela," a report to the National Assembly by Rafael Ramírez on June 9, 2005.

204 *PDVSA's fastest-growing subsidiary:* On Palmaven, see David Luhnow and Peter Millard, "As Global Demand Tightens, a Big Producer Has Own Agenda," *Wall Street Journal*, August 1, 2006.

205 *Chávez's policies were born:* Ryszard Kapuściński, *Shah of Shahs*, p. 34.

206 *"Venezuela is warming up":* The advertisement was published on December 7, 2005.

206 *And not just there:* See Juan Forero, "Chavez, Seeking Foreign Allies, Spends Billions," *New York Times*, April 4, 2006; Michael Shifter, "In Search of Hugo Chávez," *Foreign Affairs*, May 2006; and "New President

Has Bolivia Marching to Chávez's Beat," *Wall Street Journal*, May 25, 2006.

206 *Chávez made no secret:* See Jon Lee Anderson, "Fidel's Heir," in *The New Yorker*, June 23, 2008. Anderson writes, "Venezuela outspends the United States in foreign aid to the rest of Latin America by a factor of at least five. Last year, U.S. aid amounted to $1.6 billion, a third of which went to Colombia, mainly to fund Plan Colombia, a drug-eradication program administered by the U.S. security contractor DynCorp. Chávez, meanwhile, pledged $8.8 billion for the region. This included subsidized oil for Cuba, Nicaragua, and Bolivia; the purchase of public debt in Argentina; and development projects in Haiti."

207 *Caracas had a booming business:* See Sacha Feinman, "Crime and Class in Caracas," *Slate*, November 27, 2006.

207 *A highway:* Brian Ellsworth, "A Closed Bridge Mirrors Venezuela's Many Woes," *New York Times*, January 22, 2006.

208 *"magic performances, not miracles":* Fernando Coronil, *The Magical State: Nature, Money, and Modernity in Venezuela*, p. 389.

211 *For a hallucinatory period:* For a comparison of incomes, see Karl, *The Paradox of Plenty*, p. 120.

211 *"Don't study OPEC":* See Karl, *The Paradox of Plenty*, pp. xv and 4.

212 *Widespread rioting broke out:* For an overview of the Caracazo and Chávez's career, see Jon Lee Anderson, "The Revolutionary," *The New Yorker*, September 10, 2001.

213 *El Country Club has horse stables:* See Simón Romero, "Caracas Mayor Lays Claim to Golf Links to House Poor," *New York Times*, September 3, 2006.

215 *PDVSA allotted more to its social projects:* See Obiko Pearson, "Chávez's Largesse Puts Strain," Associated Press, March 27, 2007.

216 *Even during the boom years:* On price-control problems, see Peter Millard and Raúl Gallegos, "Price Caps Ail Venezuelan Economy," *Wall Street Journal*, February 15, 2006.

216 *Chávez's policies:* In a critical *Foreign Affairs* article in March/April 2008, Francisco Rodríguez, formerly the chief economist of the Venezuelan National Assembly, argued that "most health and human development indicators have shown no significant improvement beyond that which is normal in the midst of an oil boom. Indeed, some have deteriorated wor-

ryingly, and official estimates indicate that income inequality has increased."

Conclusion

219 *"the moral equivalent of war":* Carter made the remark in a speech on April 18, 1977.

220 *Riggs Bank, which helped:* Equatorial Guinea's oil revenues remain beyond the control or the regard of its citizens. According to a report published in March 2009 by Global Witness, "While the IMF has publicly reported that more than $2 billion of Equatorial Guinea's oil money is held abroad in commercial banks, it has not identified these banks. Given the history of poor management of Equatorial Guinea's oil funds, if the IMF knows where this money is, it should say so." The report posed a number of questions, including the following: "What due diligence are these commercial banks, wherever they are, doing on payments from the accounts, in order to ensure that state funds are not continuing to be diverted? Who are the signatories on the accounts?" The report is posted at www.globalwitness.org/media_library_detail.php/735/en/undue_diligence _how_banks_do_business_with_corrupt.

221 *The law is a weapon:* Siemens pleaded guilty in 2008 to violating the FCPA and agreed to pay record fines of $450 million to the Justice Department and $350 million to the Securities and Exchange Commission. Those fines were in addition to nearly 600 million euros paid to German authorities. The company had secretly made $1.4 billion in bribes across the world between 2001 and 2007. Although the American fines were record setting under the FCPA, they were not quite the result of prosecutorial zeal in America. According to a *New York Times* story on December 21, 2008 ("At Siemens, Bribery Was Just a Line Item"), German authorities began investigating Siemens in 2005 and informed American authorities a year later because Siemens shares were traded on the New York Stock Exchange. According to *Corporate Counsel* magazine, twenty-one of twenty-five new FCPA cases were self-reported in 2005–07, meaning the firms told authorities they broke the law. "Many voluntary disclosures came after violations were unearthed in the due diligence process for a merger or acquisition," noted the story, published on July 16, 2007.

221 *"There is no cure"*: Ida M. Tarbell, *The History of the Standard Oil Company*, p. 227. Shearman & Sterling, an international corporate law firm, issued a report in March 2009 that said the number of new bribery cases reached a record high in 2007, with a total of thirty-eight initiated by the Department of Justice and Securities and Exchange Commission. The figure slipped to twenty-five in 2008. The report, "FCPA Digest: Cases and Review Releases Relating to Bribes to Foreign Officials Under the Foreign Corrupt Practices Act of 1977," is posted at www.shearman.com/files/upload/LT-030509-FCPA-Digest-Cases-And-Review-Relating-to%20Bribes-to-Foreign-Officials-under-the-Foreign-Corrupt-Practices-Act.pdf. PBS's *Frontline* program "Black Money," which aired on April 7, 2009, has a Web site with extensive links to bribery-related documents from, among others, the BAE, Riggs, Halliburton and Siemens cases. See http://www.pbs.org/wgbh/pages/frontline/blackmoney/readings/.

221 *On the economic side:* On occasion, the World Bank has admitted its shortcomings. In an evaluation issued in 2007 ("World Bank Assistance to Agriculture in Sub-Saharan Africa"), the bank admitted that it should not have advised African governments to withdraw support from their farming sectors in the 1980s and 1990s. The bank wrongly expected that the private sector would provide necessary support; when that didn't happen, farmers were devastated. "In most reforming countries, the private sector did not step in to fill the vacuum when the public sector withdrew," the bank admitted. NGOs have been even more critical. In a 2003 study, "Poverty Reduction or Poverty Exacerbation? World Bank Group Support for Extractive Industries in Africa," a coalition of NGOs, including Oxfam America, Friends of the Earth-U.S. and Catholic Relief Services, noted, "The World Bank itself has produced little evidence to show that its extractive operations have contributed to poverty alleviation in sub-Saharan Africa."

222 *As I've said:* It is important to remember that dependence on a basket of natural resources or balancing resource exports with farming and industry can lead to positive outcomes, as happened in Canada, Australia and the United States. This book has focused, instead, on countries that depend to an unhealthy extent on just oil and gas.

222 *"humanity already possesses"*: See Stephen Pacala and Robert Socolow, "Stabilization Wedges: Solving the Climate Problem for the Next 50 Years with Current Technologies," *Science Magazine*, August 13, 2004.

223 *"a limited set of monumental tasks"*: Robert Socolow, in a keynote speech, "Stabilization Wedges: Mitigation Tools for the Next Half-Century," at a climate-change symposium in Exeter, United Kingdom, Feburary 1–3, 2005.

224 *The battalion had lost:* See Peter Maass, "Good Kills," *New York Times Magazine*, April 20, 2003.

BIBLIOGRAPHY

Baer, Robert. *Sleeping with the Devil: How Washington Sold Our Soul for Saudi Crude.* New York: Crown, 2003.

Bakan, Joel. *The Corporation: The Pathological Pursuit of Profit and Power.* New York: Free Press, 2004.

Barthes, Roland. *Camera Lucida: Reflections on Photography.* New York: Hill and Wang, 1981.

Bronson, Rachel. *Thicker Than Oil: America's Uneasy Partnership with Saudi Arabia.* New York: Oxford University Press, 2006.

Bryce, Robert. *Cronies: Oil, the Bushes, and the Rise of Texas, America's Superstate.* New York: Public Affairs, 2004.

————. *Gusher of Lies: The Dangerous Delusions of "Energy Independence."* New York: Public Affairs, 2008.

Bumiller, Elisabeth. *Condoleezza Rice: An American Life.* New York: Random House, 2007.

Bush, George W. *A Charge to Keep.* New York: Morrow, 1999.

Bush, George, and Brent Scowcroft. *A World Transformed.* New York: Vintage Books, 1999.

Cockburn, Patrick. *Muqtada: Muqtada al-Sadr, the Shia Revival, and the Struggle for Iraq.* New York: Scribner, 2008.

Coll, Steve. *The Bin Ladens: An Arabian Family in the American Century.* New York: Penguin Press, 2008.

————. *Ghost Wars: The Secret History of the CIA, Afghanistan, and bin Laden, from the Soviet Invasion to September 10, 2001.* New York: Penguin Press, 2004.

Collier, Paul. *The Bottom Billion: Why the Poorest Countries Are Failing and What Can Be Done About It.* New York: Oxford University Press, 2007.

Coronil, Fernando. *The Magical State: Nature, Money, and Modernity in Venezuela.* Chicago: University of Chicago Press, 1997.

Deffeyes, Kenneth S. *Beyond Oil: The View from Hubbert's Peak.* New York: Hill and Wang, 2005.

———. *Hubbert's Peak: The Impending World Oil Shortage.* Princeton: Princeton University Press, 2001.

Freeland, Chrystia. *Sale of the Century: Russia's Wild Ride from Communism to Capitalism.* New York: Crown, 2000.

Gaidar, Yegor. *Collapse of an Empire: Lessons for Modern Russia.* Washington, D.C.: Brookings Institution Press, 2007.

Gelbspan, Ross. *Boiling Point: How Politicians, Big Oil and Coal, Journalists, and Activists Have Fueled the Climate Crisis—and What We Can Do to Avert Disaster.* New York: Basic Books, 2004.

Ghazvinian, John. *Untapped: The Scramble for Africa's Oil.* New York: Harcourt, 2007.

Goldman, Marshall I. *Petrostate: Putin, Power and the New Russia.* New York: Oxford University Press, 2008.

Goltz, Thomas. *Azerbaijan Diary: A Rogue Reporter's Adventures in an Oil-Rich, War-Torn, Post-Soviet Republic.* Armonk, N.Y.: M.E. Sharpe, 1999.

Goodell, Jeff. *Big Coal: The Dirty Secret Behind America's Energy Future.* New York: Houghton Mifflin, 2006.

Goodstein, David. *Out of Gas: The End of the Age of Oil.* New York: Norton, 2004.

Harris, David. *The Crisis: The President, the Prophet, and the Shah—1979 and the Coming of Militant Islam.* New York: Little, Brown, 2004.

Hiltermann, Joost R. *A Poisonous Affair: America, Iraq and the Gassing of Halabja.* New York: Cambridge University Press, 2007.

Hoffman, David E. *The Oligarchs: Wealth and Power in the New Russia.* New York: Public Affairs, 2002.

Humphreys, Macartan, Jeffrey D. Sachs, and Joseph E. Stiglitz, eds. *Escaping the Resource Curse.* New York: Columbia University Press, 2007.

Jamieson, Kathleen Hall, and Paul Waldman. *The Press Effect: Politicians, Journalists, and the Stories That Shape the Political World.* New York: Oxford University Press, 2004.

Kane, Joe. *Savages.* New York: Alfred A. Knopf, 1995.

Kapuściński, Ryszard. *Shah of Shahs.* New York: Vintage International, 1992.

Karl, Terry Lynn. *The Paradox of Plenty: Oil Booms and Petro-States.* Berkeley: University of California Press, 1997.

Kinzer, Stephen. *All the Shah's Men: An American Coup and the Roots of Middle East Terror.* Hoboken, N.J.: John Wiley, 2003.

Klebnikov, Paul. *Godfather of the Kremlin: The Decline of Russia in the Age of Gangster Capitalism.* New York: Harvest Books, 2001.

Kleveman, Lutz. *The New Great Game: Blood and Oil in Central Asia.* New York: Atlantic Monthly, 2003.

Klitgaard, Robert. *Tropical Gangsters: One Man's Experience with Development and Decadence in Deepest Africa.* New York: Basic Books, 1990.

Kolbert, Elizabeth. *Field Notes from a Catastrophe: Man, Nature, and Climate Change.* New York: Bloomsbury, 2006.

Kotkin, Stephen. *Armageddon Averted: The Soviet Collapse, 1970–2000.* New York: Oxford University Press, 2001.

LeVine, Steve. *The Oil and the Glory: The Pursuit of Empire and Fortune on the Caspian Sea.* New York: Random House, 2007.

Liniger-Goumaz, Max. *A l'Aune de la Guinée équatoriale.* Geneva: Les Editions du Temps, 2003.

Maier, Karl. *This House Has Fallen: Midnight in Nigeria.* New York: Public Affairs, 2000.

Mann, James. *Rise of the Vulcans: The History of Bush's War Cabinet.* New York: Penguin Books, 2004.

Margonelli, Lisa. *Oil on the Brain: Adventures from the Pump to the Pipeline.* New York: Nan A. Talese/Doubleday, 2007.

McKibben, Bill. *The End of Nature.* New York: Random House, 1989.

Milgram, Stanley. *Obedience to Authority.* 1974. Reprint, New York: Harper Perennial, 1983.

Munif, Abdelrahman. *Cities of Salt.* New York: Vintage International, 1989.

Oliveira, Ricardo Soares de. *Oil and Politics in the Gulf of Guinea.* New York: Columbia University Press, 2007.

Philips, Kevin. *American Theocracy: The Peril and Politics of Radical Religion, Oil, and Borrowed Money in the 21st Century.* New York: Viking, 2006.

Pollack, Kenneth M. *The Persian Puzzle: The Conflict Between Iran and America.* New York: Random House, 2005.

Rasheed, Madawi al-. *A History of Saudi Arabia.* Cambridge: Cambridge University Press, 2002.

Roberts, Adam. *The Wonga Coup: Guns, Thugs and a Ruthless Determination to Create Mayhem in an Oil-Rich Corner of Africa*. New York: Public Affairs, 2006.

Roberts, Paul. *The End of Oil: On the Edge of a Perilous New World*. New York: Houghton Mifflin, 2004.

Sampson, Anthony. *The Seven Sisters: The Great Oil Companies and the World They Shaped*. New York: Viking Press, 1975.

Saro-Wiwa, Ken. *A Month and a Day: A Detention Diary*. New York: Penguin Books, 1995.

Sawyer, Suzana. *Crude Chronicles: Indigenous Politics, Multinational Oil, and Neoliberalism in Ecuador*. Durham, N.C.: Duke University Press, 2004.

Schweizer, Peter. *Victory: The Reagan Administration's Secret Strategy That Hastened the Collapse of the Soviet Union*. New York: Atlantic Monthly, 1996.

Simmons, Matthew R. *Twilight in the Desert: The Coming Saudi Oil Shock and the World Economy*. Hoboken, N.J.: John Wiley, 2005.

Sinclair, Upton. *Oil!* 1927. Reprint, Berkeley: University of California Press, 1977.

Speth, James Gustave. *The Bridge at the Edge of the World: Capitalism, the Environment, and Crossing from Crisis to Sustainability*. New Haven: Yale University Press, 2008.

Stegner, Wallace. *Discovery! The Search for Arabian Oil*. Beirut: Middle East Export Press, 1971.

Tarbell, Ida M. *The History of the Standard Oil Company*. 1904. Reprint, Mineola, N.Y.: Dover Publications, 2003.

Tertzakian, Peter. *A Thousand Barrels a Second: The Coming Oil Break Point and the Challenges Facing an Energy Dependent World*. New York: McGraw-Hill, 2006.

Theroux, Peter. *Sandstorm: Days and Nights in Arabia*. New York: Norton, 1990.

Trofimov, Yaroslav. *The Siege of Mecca: The 1979 Uprising at Islam's Holiest Shrine*. New York: Anchor Books, 2008.

Vitalis, Robert. *America's Kingdom: Mythmaking on the Saudi Oil Frontier*. New York: Verso, 2009.

Wright, Lawrence. *The Looming Tower: Al-Qaeda and the Road to 9/11*. New York: Alfred A. Knopf, 2007.

Yeomans, Matthew. *Oil: Anatomy of an Industry*. New York: New Press, 2004.

Yergin, Daniel. *The Prize: The Epic Quest for Oil, Money and Power*. New York: Simon & Schuster, 1991.

INDEX